ATLAS OF CELL BIOLOGY

L

ATLAS OF CELL BIOLOGY

Jean-Claude Roland, D.Sc.
Professeur,
Université Paris VI

Annette Szöllösi, D.Sc.
Maître-Assistant,
Université Pierre et Marie Curie

Daniel Szöllösi, Ph.D.
Professeur associé,
Université Paris-Sud

Translated from the French by
Salvatore F. Vitale, Ph.D.

With additional material by
James D. Jamieson, M.D.
Yale University School of Medicine

Little, Brown and Company
Boston

Suggested Reading list prepared by Anthony Shields, Massachusetts Institute of Technology.

Library of Congress Catalog Card No. 76-56028

ISBN 0-316-75450-1

Printed in the United States of America

PREFACE

This atlas is directed at students of introductory cell biology. Students do have at their disposal synopses and manuals dealing with fundamental aspects of this discipline. Nevertheless, several years of teaching have convinced us of the usefulness of collecting in a concise form material appropriate for a basic course of lectures and laboratory work.

An atlas grouping photomicrographs and simple functional diagrams could link these forms of teaching and guide the reader to associate ideas of structure and function. Comparison of results obtained by applying different methods to the same biological material could furnish the basis for an approach to certain problems in methodology.

Colleagues at home and abroad have kindly sent us their records and results. A difficult problem of choice arose since we desired to limit ourselves to essential data, presented in a simple manner in a work of restricted size.

Our aim will have been achieved if this atlas arouses the student's interest in the study of the cell and helps in the understanding of its organization.

J.-C. R.
A. S.
D. S.

HISTORICAL NOTE

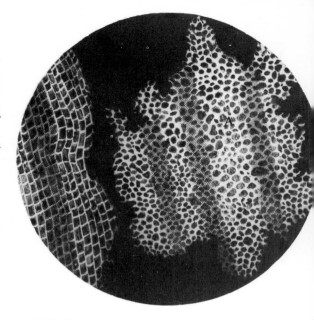

The conception of the organism as an association of living individuals or units gained acceptance only slowly during the historical course of Biology. The organization of the cell and its contents, and the identification and significance of the structures contained, were demonstrated in gradual succession, emerging from a multitude of contradictory or erroneous concepts. Progress in research met with numerous difficulties that were the result of the very fragility of cellular matter, as well as the finesse of the methods used in their study.

1662: First Representation of Cells. This figure is taken from the chapter "Concerning the texture of cork and of cells or pores of some spongy bodies of the same kind" in *Micrographia,* the renowned work of Robert Hooke. Over the course of nearly two centuries, little further knowledge of the microscopic structure of living beings accrued (it was only in 1831 that an "aureole" or "nucleus" in plants was mentioned by Robert Brown).

1830–1860: Development of Cellular Theory. The figure is taken from *Microscopic Investigations on the Correspondence between Structure and Growth of Animals and Plants,* by Theodor Schwann (1839). Organisms are interpreted as "collective beings," the units of which are cells. "These units are also living individuals, each enjoying the properties of growth, of reproduction, and of altering itself within certain limits" (Brisseau de Mirbel, 1839). Schwann thought that cells arose from inorganic "fluid," but it was soon apparent that spontaneous generation of cells could not be demonstrated and that all cells derive from a preexisting cell (Rudolf Virchow, 1858). Thanks to improvements in the light microscope (Abbe's microscope, 1878), there was a succession of discoveries: mitosis, 1870–1880; mitochondria, 1890–1900; Golgi apparatus, 1898–1900.

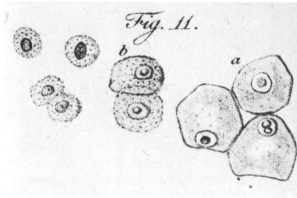

1945–the present: Introduction of Electron Microscopy to the Study of Biology. This is a reproduction of one of the first electron micrographs of cellular ultrastructure, published by A. Claude, K. Porter, and E. Pickels in 1947. It is an image of a suspended cell (chicken blood macrophage), fixed by osmium tetroxide and deposited on the specimen holder of the microscope. The thin areas at the border of the cell allow one to recognize the outline of mitochondria and portions of the endoplasmic reticulum. The nucleus (lower left) is wholly opaque to the electrons.

In the years that followed, methods improved. Embedding specimens in plastic made it possible to obtain ultrathin sections; suitable cytochemical and cytoenzymological methods were developed; freeze-etching was introduced in 1963; electron microscopes were diversified (high voltage, microprobe, scanning electron microscope). At the same time, biochemical methods were improving and could be applied to isolated cellular organisms. The development of autoradiography made it possible to localize labeled molecules within the cell. Meanwhile, the rapidly expanding field of immunology was aided by the visualization of antibodies tagged with fluorescent dyes or ferritin.

Cellular biology was now a vigorous science.

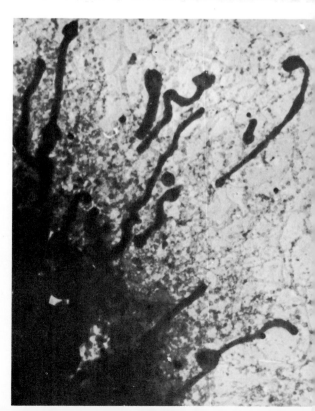

CONTENTS

MAGNIFICATIONS

1 meter $= 10^3$ mm (millimeters)
$\quad\quad\quad\; = 10^6\,\mu$ (microns)
$\quad\quad\quad\; = 10^9$ nm (nanometers)
$\quad\quad\quad\; = 10^9\,m\mu$ (millimicrons)
$\quad\quad\quad\; = 10^{10}$ Å (angstroms)

Examples of Scales Used:

 1 μm $= \times\ 25,000$

0.5 μm $= \times\ 25,000$

0.1 μm

 100 nm $= \times\ 250,000$

1,000 Å

Magnifications are given in micrometers, nanometers, and angstroms. Except where indicated, the scale line on the micrographs represents 1 micrometer.

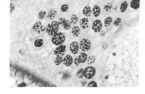

Epithelial tissue
$\times\ 250$

Chromosomes
$\times\ 1,000$

Mitochondrion
$\times\ 10,000$

Nuclear envelope
$\times\ 22,000$

Isolated mitochondrial DNA
$\times\ 50,000$

Virus
$\times\ 1,000,000$

Protein crystal
$\times\ 210,000$

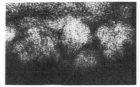

Cytoplasmic membrane
$\times\ 270,000$

Membrane subunits
$\times\ 700,000$

Enzymatic complex
$\times\ 1,000,000$

ATLAS OF CELL BIOLOGY

INTRODUCTION: FROM STRUCTURE TO ULTRASTRUCTURE

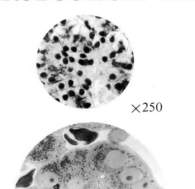

×250

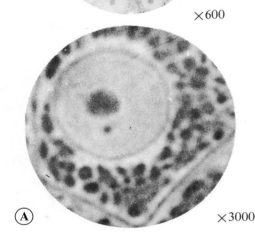

×600

Observation of living cells with the light microscope and of fixed and sectioned cells by electron microscopy have now been closely correlated. The former has the advantage that it produces dynamic images of low resolution while the latter reveals static images of the cell's fine structure at high resolution. Bridging the gap between the two optical tools depended on cell fractionation and biochemical studies, as described later.

On the opposite page are intact, unstained, living cells seen by phase-contrast light microscopy. The shape and random arrangement of cells, nuclei, nucleoli, and cytoplasmic granules are readily seen. In vitro culture of a mouse tumor, ×1000.

×1000

(A)

×3000

×5000

A. **Fixed and stained cells by light microscopy.** Natural light.

The tissue (frog liver in this case) was embedded in epoxy and sectioned in order to obtain a thin preparation. With light microscopy the theoretical resolving power of 0.2 μm is such that usable direct magnifications are limited to ×1000 to ×2000. In addition to the nuclei and nucleoli, one can identify mitochondria and cytoplasmic membrane.

B. **Fixed, contrasted cells by electron microscopy.**

Ultrathin sections (500 Å to 800 Å thick) are examined. The tissue (frog liver) was embedded in plastic before sectioning in order to obtain an ultrathin preparation. The resolving power of electron microscopy is 3 Å to 10 Å, which allows the observation of much finer detail and permits the use of high magnifications (greater than ×100,000).

The magnification indicated throughout is the total magnification (direct magnification by the microscope × photographic magnification).

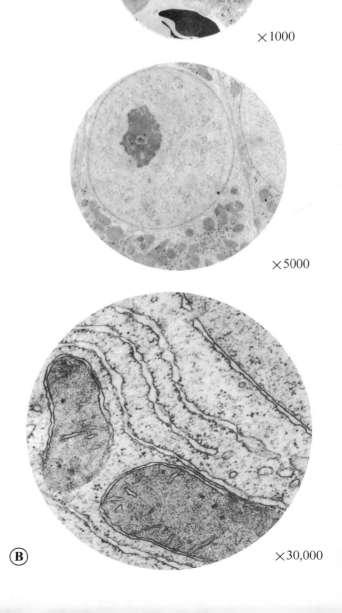

(B)

×30,000

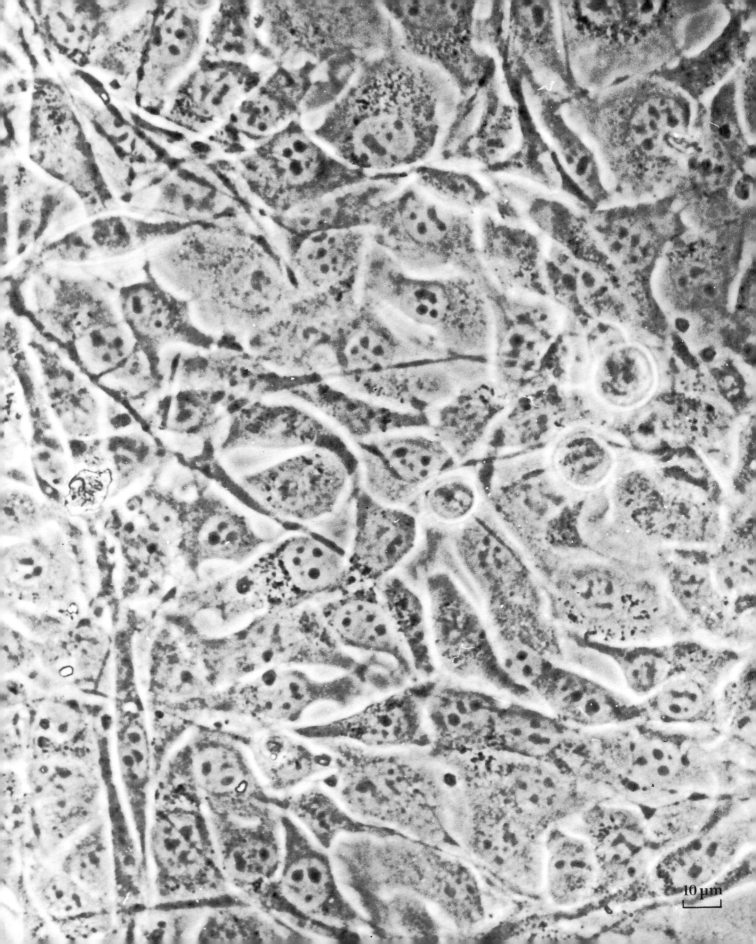

10 μm

1. GENERAL ORGANIZATION OF THE CELL

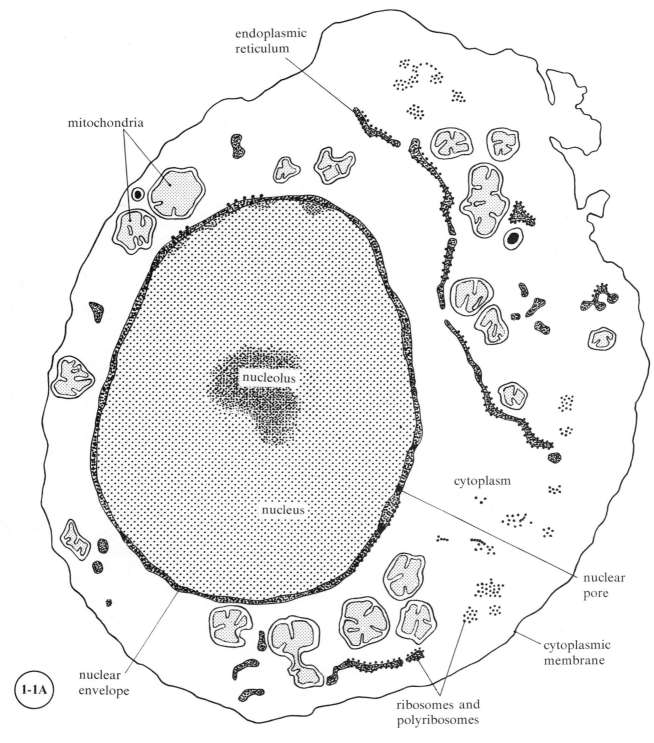

1-1. **General view of a cell showing simple organization.** This cell, an oogonium, comes from a mouse embryo. It is as yet only slightly differentiated. ×11,000.

The nucleus represents a cellular territory limited by the nuclear envelope. It contains some dispersed chromatin and a denser nucleolus. The cytoplasmic territory is compartmented. It contains mitochondria and endoplasmic reticulum. Numerous ribosomes are associated with the reticular membrane or are dispersed in the cell sap. They may occur as isolated units or may be grouped in polyribosomes.

The cytoplasmic membrane ensures contact with the medium.

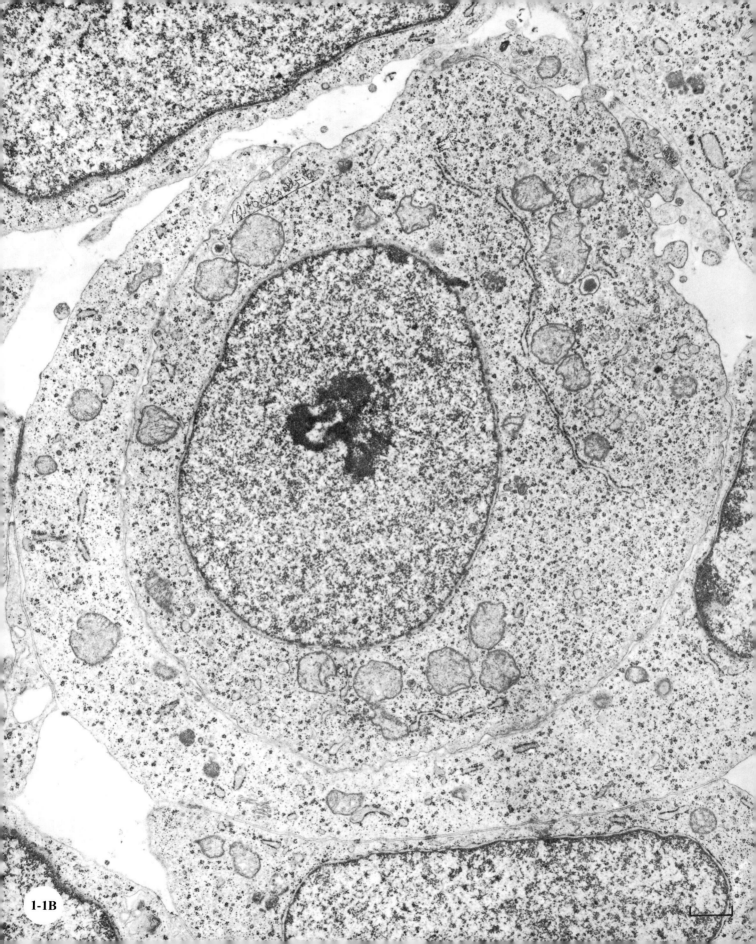

1-1B

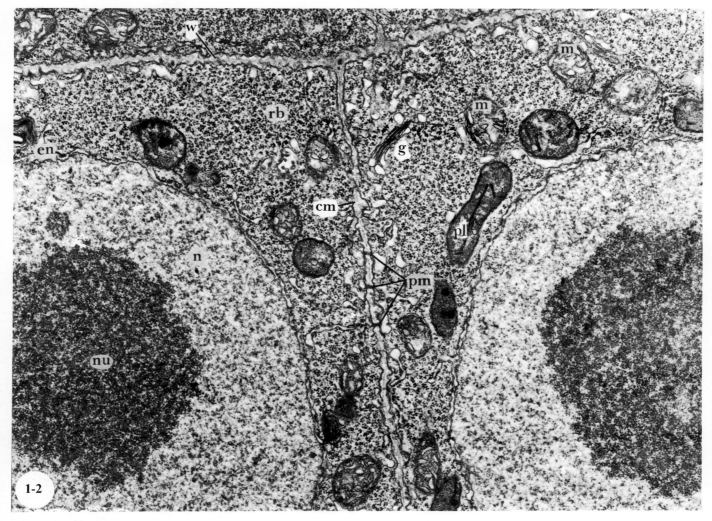

1-2

1-2. **Young plant cell.** Embryo of *Capsella bursa-pastoris* at the eight-cell stage, ×25,000. (R. Schultz and W. Jensen.) The following can be identified: the nucleus (n); the very voluminous nucleolus (nu); the ribosomes abounding in the cytoplasm (rb); the mitochondria in the cytoplasm (m); a dictyosome representing the Golgi apparatus (g); the nuclear envelope (en); and the cytoplasmic membrane (cm). Two constituents are characteristic of plants: the intercellular wall (w) and the plastids (pl), which have as yet little structure. Cells communicate by cytoplasmic bridges, the plasmodesmata (pm).

1-3. **Differentiating animal cell.** Blood cell from a mouse embryo, ×20,000.
The nucleus is lobed, and the nucleocytoplasmic ratio is elevated. Chromatin (ch) is condensed in hetero-chromatic masses. Note the numerous pores (np) of the nuclear envelope (en). Golgi apparatus (g) is represented in the cytoplasm by parallel cisternae and vesicles in the vicinity of a pair of centrioles, C_1 and C_2. Besides ribosomes and polyribosomes, the cytoplasm contains mitochondria (m), endoplasmic reticulum (er), and phagocytic vacuoles (phv). The profile of the cytoplasmic membrane (cm) is sinuous.
Figures 1-4 and 1-5 on the following pages show a differentiated cell as observed by two complementary techniques, chemical fixation and cryofixation. The tissue was rat liver. (L. Orci, A. Matter, and C. Rouiller.)

1-4. **Chemical fixation.** The tissue was fixed with glutaraldehyde and osmium tetroxide and embedded in plastic. Ultrathin sectioning and positive staining were then done. ×36,000.

1-5. **Cryofixation** at −150°C was done; then the tissue was freeze-fractured. The direction of shadowing is indicated by an *arrow.* ×23,000.
Note that the two layers of the nuclear envelope are particularly well seen after freeze-fracture. n = nucleus; np = nuclear pore; en = nuclear envelope; gl = glycogen; er = endoplasmic reticulum; m = mitochondria; cm = cytoplasmic membrane; d = dictyosome; bc = bile canaliculus.

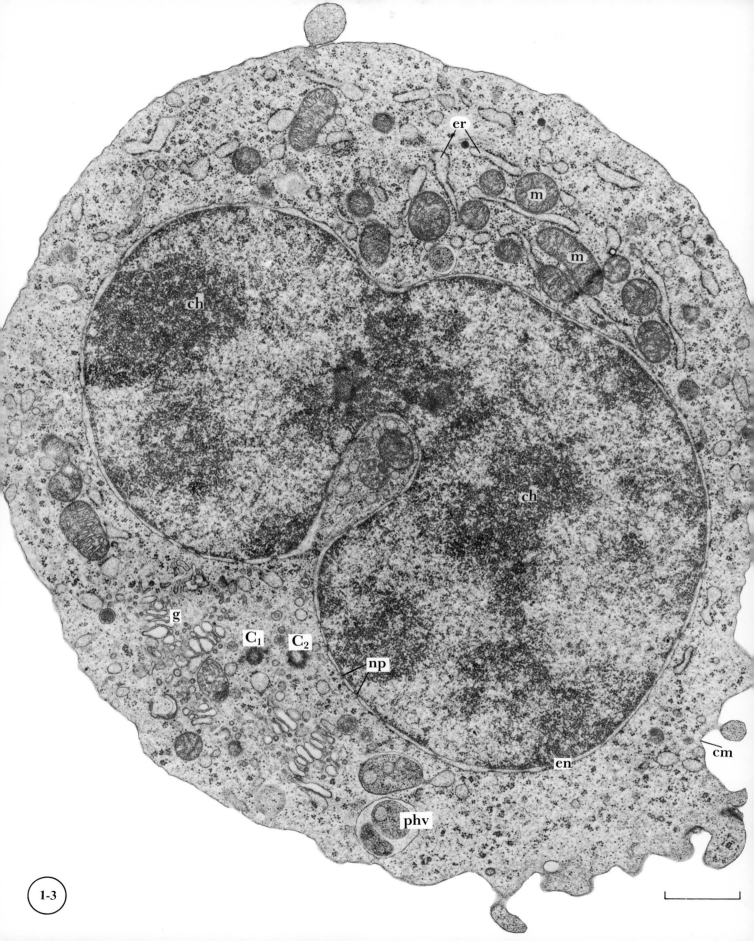

1-3

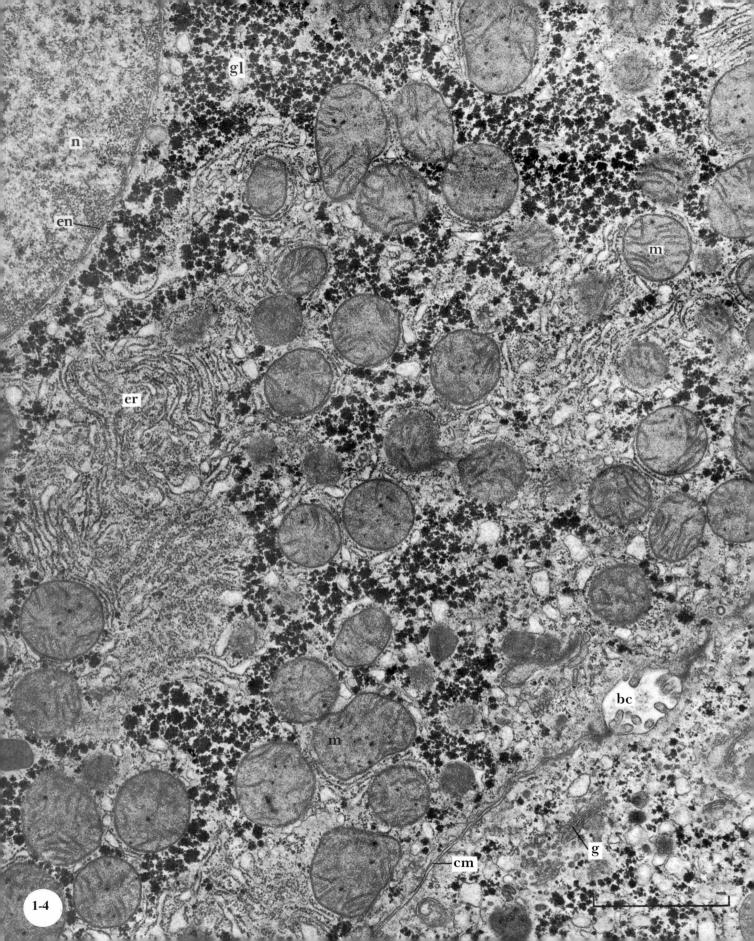

1-5

2. CELLULAR FRACTIONATION

The different cellular compartments can be separated for the purpose of analysis (Fig. 2-1). Mechanical grinding produces a cell homogenate (A), within which are dispersed the organelles. A first differential centrifugation (B) at increasing g force separates organelles into groups of decreasing size or density which are subsequently purified by discontinuous or continuous density gradient centrifugation (C). Under certain cell fractionation conditions, organelles retain their structural and functional integrity. Refinements have been added to the method, and one can actually dissociate the components of the organelles themselves and make them function separately in vitro. Density gradient centrifugation is the method of choice for the biochemical and physiologic study of cellular functioning. Control of the purity of fractions and of the integrity of organelles by electron microscopy and biochemistry permits associating structure and function.

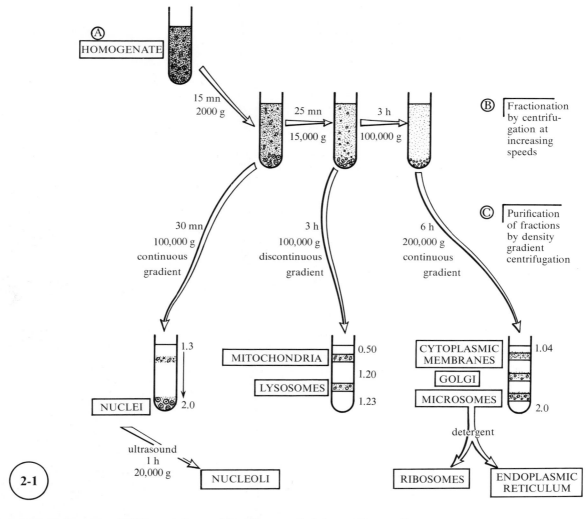

2-1. **Simplified diagram of differential centrifugation applied to rat liver cells.**
Tissue homogenization and all centrifugations are made at 4°C in buffered sucrose solutions. Solutions of different molarities are used to make up gradients in which organelles are partitioned as a function of their density. The numbers indicate the molarities of the sucrose solutions. mn = minute; h = hour; g = gravity unit.

2-2. **Control by electron microscopy of the isolated and purified fractions derived from rat liver cells.**
A. Microsomal fraction, ×42,500. (M. Wibo et al.)
B. Golgi fraction, ×30,000. (D.-J. Morré.)
C. Mitochondrial fraction, ×27,000. (J. André, R. Toury, and M. Levy.)
D. Nuclear fraction, ×4500. (P. Mentré.)

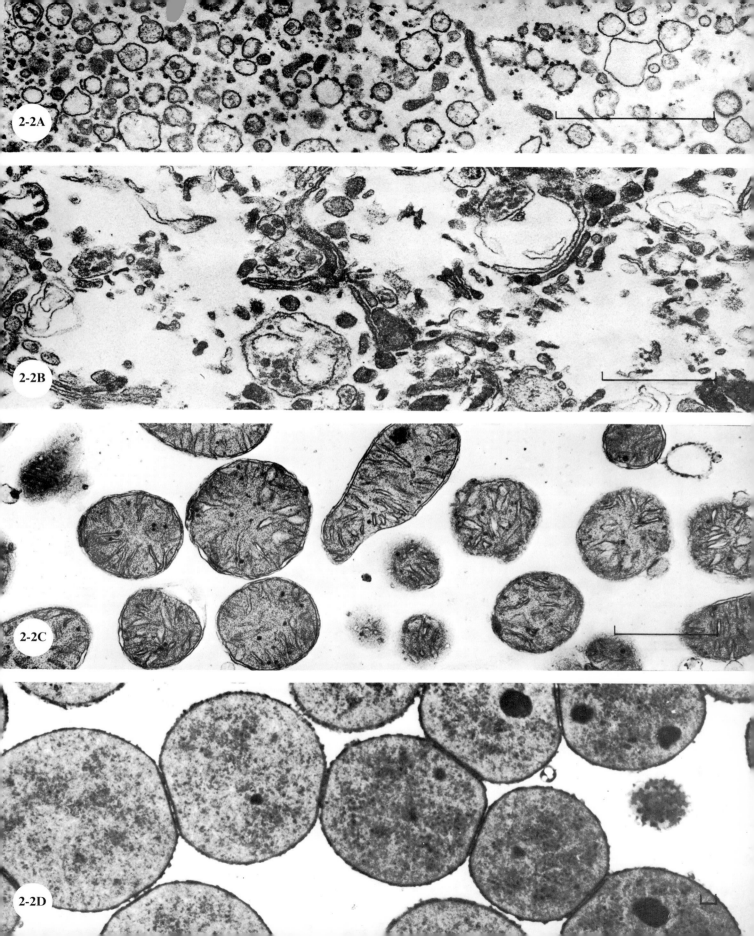

3. CYTOPLASMIC MEMBRANE

A biologic membrane is an active, coordinated assembly of molecules capable of ensuring exchanges of materials with its environment. The cytoplasmic membrane, or plasmalemma, establishes cell contact with the external environment. It serves in a number of ways, including protection, adhesiveness, reception of exogenous information, and selective and oriented transfer of metabolites. Essentially, it is constituted of proteins and phospholipids. Several models have been proposed to explain the structural and functional properties of the membrane, but none is entirely satisfactory. In fact, membranes are currently the focus of a great deal of research. Figures 3-1, 3-2, and 3-3 show features of the cytoplasmic membrane as seen in pea root tissue.

3-1. Tripartite structure. ×300,000.

After chemical fixation, sectioning, and positive staining, the cytoplasmic membrane appears to be made up of two dense layers that are separated by a clear space. The two layers are often asymmetrical, which could reflect functional polarity of the membrane.

3-2. Particulate structure. ×110,000.

Particulate structure is revealed reproducibly by freeze-fracture. The number of particles can be in excess of 3000 per square micrometer, each particle having a diameter of about 100 Å. The distribution of particles varies with cellular activity. Particles are probably proteins.

3-3. Subunits of membranes. ×1,200,000.

At very high magnification and after negative staining subunits may be seen on ultrathin frozen sections. Some of these subunits could correspond with multienzymatic complexes (*arrow*).

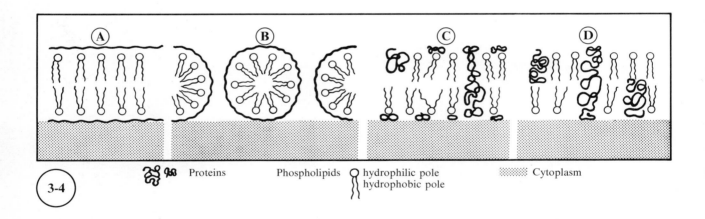

Proteins   Phospholipids   hydrophilic pole / hydrophobic pole   Cytoplasm

3-4

3-4. Molecular architecture of the membrane.

Numerous models of the association of proteins with polar lipids (phospholipids) and nonpolar lipids have been proposed to account for membrane properties. In one, the membrane appears as a regular, bilamellar structure (A); and in another, as a micellar structure (B). Currently, membranes are considered to be structures consisting of proteins embedded in a fluid matrix of lipids. The proteins (C, D) are loosely associated with the lipid bilayer; they are either partly embedded in it, or span it entirely. The proteins are active transporters, multienzymatic complexes, hormonal receptors, and so on.

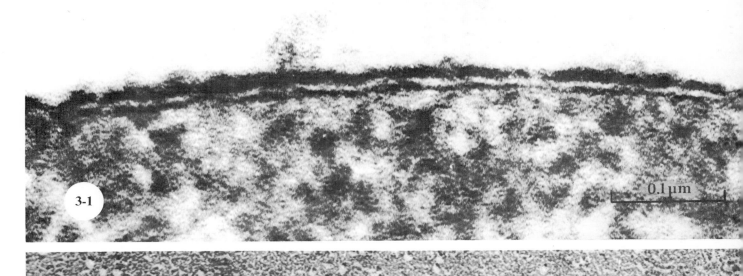

3-1

0.1 μm

3-2

0.1 μm

3-3

0.01 μm

MORPHOLOGIC CHARACTERISTICS

In most cells, the external surface of the membrane is covered by acid polysaccharides and glycoproteins, many of which are negatively charged because of sialic acid and sulfated saccharides. In some cells, the glycoproteins are enormous and protrude extensively into the aqueous environment. This forms a cell coat often termed the *glycocalyx*. The cell membrane is specialized structurally and functionally, examples being microvilli and intercellular junctions. Membrane and cell coat can be identified locally as microvilli on cellular surfaces, in contact with the external medium; and as junctions and intercellular cement between cells in contact in a tissue. Junctions have various structures and functions. They have a mechanical role in the attachment system. Some act in cellular exchanges by modifying the electric potential of membranes.

3-5. Microvilli. Rat oocyte, ×56,000.
These are finger-like evaginations of the membrane. They contain parallel microfilaments arranged in a circle (*arrow*). cm = cytoplasmic membrane; mv = microvilli.

3-6. Glycocalyx. Medusa ovum, ×15,000.
Examination of a thick section of this cell was made possible by using a very high voltage microscope (see p. 113). It shows the branched nature of microvilli (mv) and the rich felt-work of polysaccharides which covers the microvilli and forms the glycocalyx (gx).

3-7. Intercellular junction. Mouse embryo, ×72,000.
This junction provides a very tight contact between two cells and allows exchanges of ions. It is therefore called an electronic junction or gap junction (*arrow*). cm = cytoplasmic membrane.

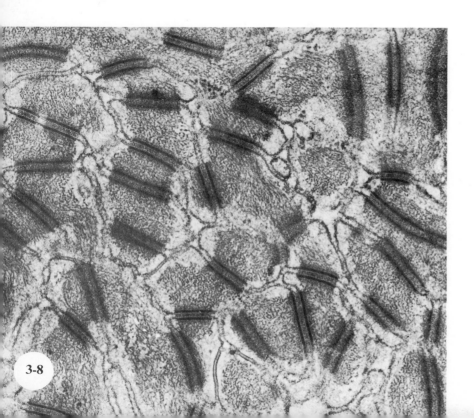

3-8

3-8. Desmosome. Calf epidermis, ×50,000. (R. Leloup, L. Laurent, and P. Drochmans.) Attachment occurs between cells at this junction. The membranes of the two cells are woven together locally by a dense material and by numerous filaments. The abundance of desmosomes in epidermis explains the strong cohesion of this tissue.

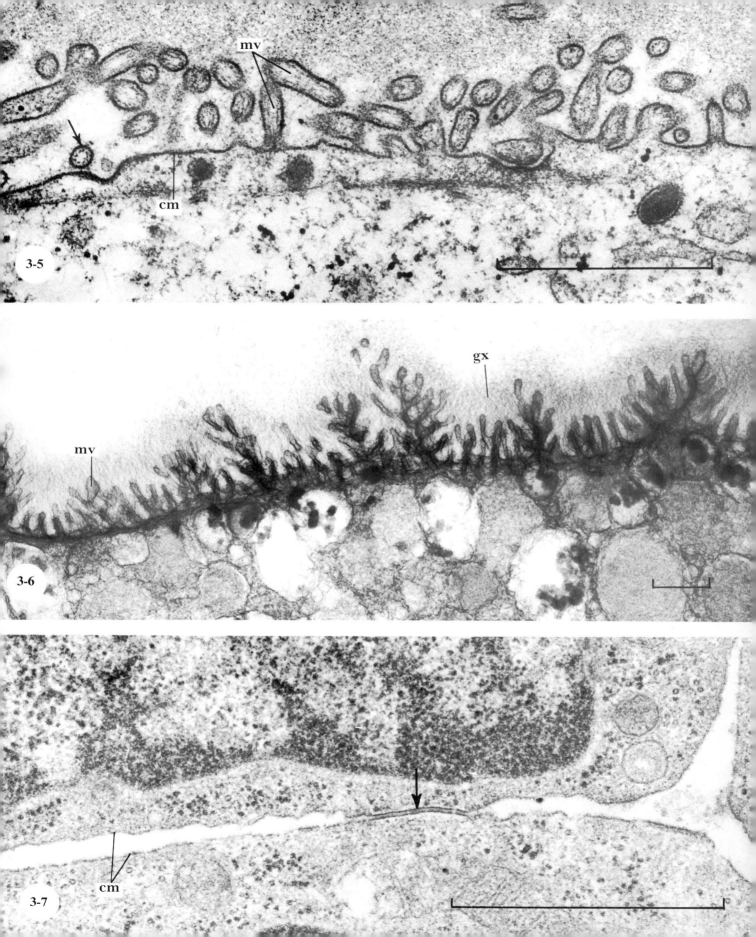

PHYSIOLOGIC ACTIVITIES

Of the exchanges which occur at the membrane, perhaps the most characteristic is active transport. This is performed by proteins, T, which transfer substances, S, from one compartment to the other across the membrane. Energy is required for active transport, and this is provided by ATP (Fig. 3-9; see also p. 38). Other membrane components act as receptors for hormones. Within a given immune system, some of these surface molecules are antigens.

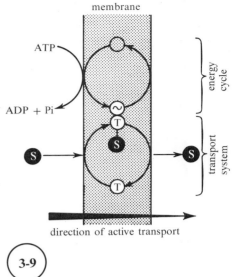

3-9. Active transport system. Transport of substances against a concentration gradient can occur through use of energy provided by a coupled energy cycle.

3-10. Membrane enzymatic activity. Conducting cell of elder leaf, ×20,000.
ATP was provided for the cell. It was specifically hydrolyzed by an ATPase at the cytoplasmic membrane (cm) according to the reaction:

$$ATP \longrightarrow ADP + Pi$$

Liberated inorganic phosphate (Pi) may be shown in situ by cytochemistry (see p. 108). Apart from the membrane (cm) other cellular structures show no activity.
n = nucleus; nu = nucleolus; m = mitochondria.

3-11. Antigenic activity of the membrane. Human red blood cell, ×120,000. (R. Lee and J. Feldmann.)
Membrane glycoproteins and polysaccharides are antigens in immune reactions. They can be detected by immunocytochemistry because of the specificity of the antibody-antigen reaction. Blood group A is the example in this case.

3-12. Synaptic junction. Nerve cells, ×100,000. (J. Taxi.)
Conduction of nerve impulses is an extremely rapid process that activates both electrical phenomena along the membrane and chemical substances (acetylcholine or norepinephrine) at the synapse. The synapse is the zone of contact between two nerve cells. Emission of active substances causes the transfer of the impulse from one cell to another. The photograph shows the endings of fiber A in contact with fiber B. The synaptic vesicles of fiber A contain acetylcholine, which is discharged into the intercellular space, thus exciting membrane receptors of fiber B. Thus the cytoplasmic membranes (cm) of a synaptic junction are adapted to a special function.

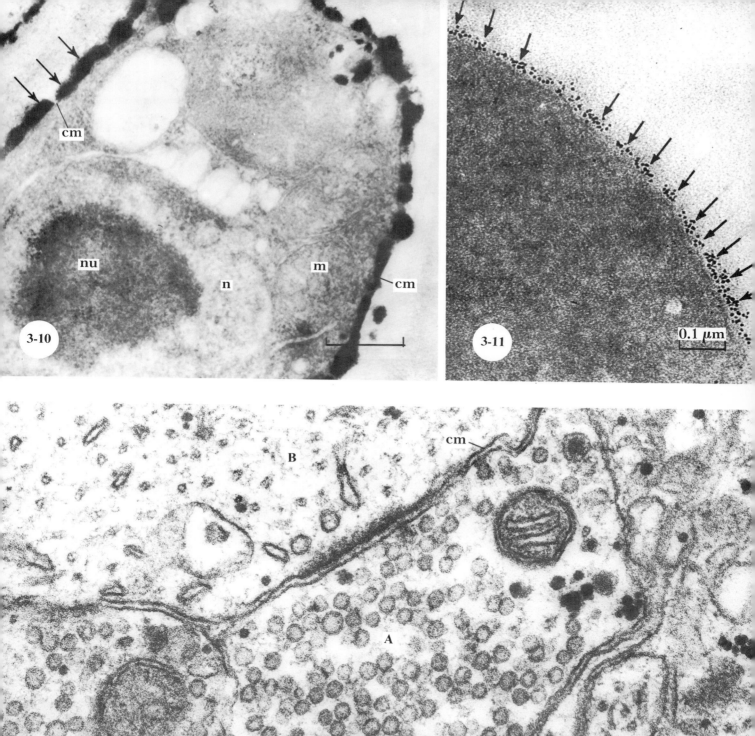

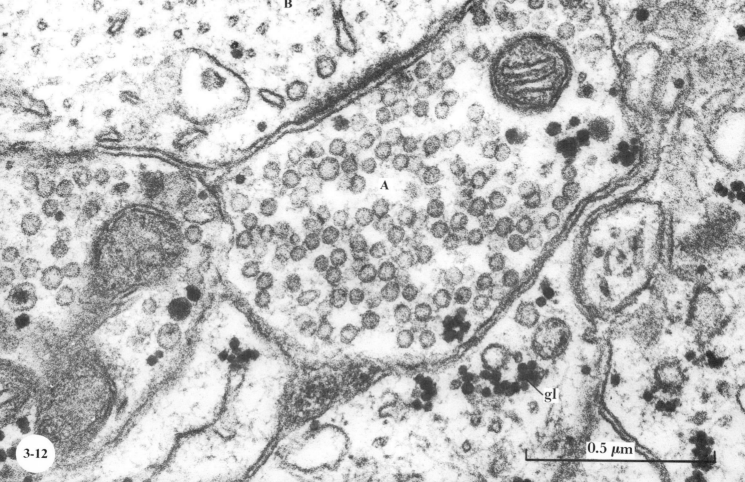

4. INTERNAL MEMBRANE SYSTEM

Continuity can be shown between certain cytoplasmic compartments delimited by membranes: rough or smooth endoplasmic reticulum, Golgi apparatus, lysosomes, vacuoles, and such. This continuity has to be envisaged not only in space but also in time: membrane flow produces a transformation and progressive movement from one compartment to another. Such components thus seem to be differentiated and specialized regions of an internal membrane system which is being renewed constantly.

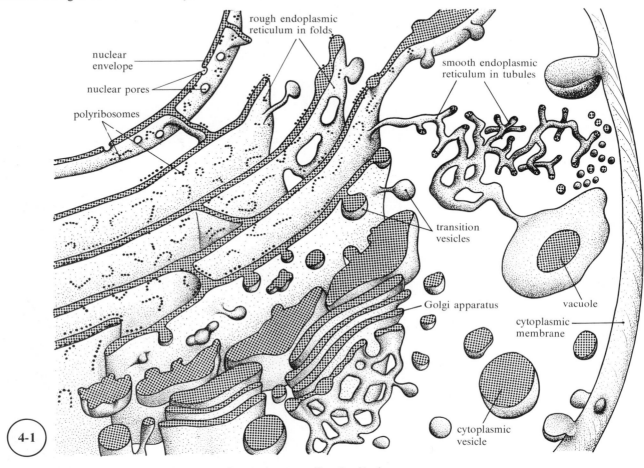

4-1. Compartments of the internal membrane system: Continuity in space.
Isolated components such as cytoplasmic vesicles cannot be linked directly to the internal membrane system except by a stepwise study which specifies their origin and the import of their migration. Thus, static images which show vesicles opening to the exterior in relation to the cytoplasmic membrane do not permit a distinction between exocytosis (secretion) and endocytosis (absorption).

4-2. Internal membrane system. Follicular cell of rat ovary, ×34,000.
Some ribosomes are on the external surface of the membrane of the nuclear envelope. Polyribosomes (ps) are numerous in the cytosol (cell sap). en = nuclear envelope; g = Golgi apparatus, transverse and tangential sections of which are seen; cm = cytoplasmic membrane; rer = rough endoplasmic reticulum; ser = smooth endoplasmic reticulum; ly = lysosome.

4-3. Two views of the endoplasmic reticulum.
A. Folds of rough endoplasmic reticulum. Frog pancreas, ×25,000.
B. Tubules of smooth endoplasmic reticulum. Frog liver, ×25,000. This type of reticulum is especially adapted for the synthesis and transport of lipoproteins and steroids.

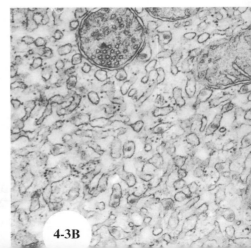
4-3B

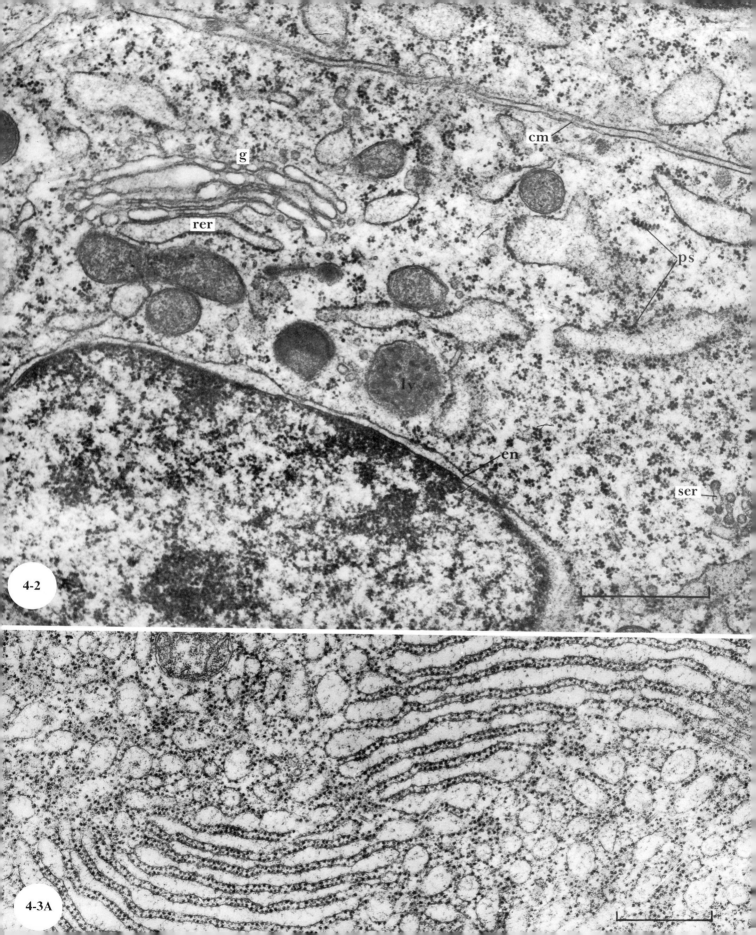

ENDOPLASMIC RETICULUM AND RIBOSOMES: PROTEIN SYNTHESIS

Endoplasmic reticulum is a cellular transport system. It is active in lipid biosynthesis. In contractile cells it concentrates cations in quantities which are sometimes considerable. Its essential role lies in the transport and synthesis of proteins on the polysome and their segregation into its external space.

4-4. Cell actively synthesizing proteins. Follicular cell of rat ovary, ×27,500. en = nuclear envelope; cm = cytoplasmic membrane; rer = rough endoplasmic reticulum, the cisternae of which are dilated and contain a dense product, glycoproteins.

4-5. Detail of the rough endoplasmic reticulum cut tangentially. ×180,000.
ps = polyribosomes, which are borne on the external surface of membranes of the endoplasmic reticulum. The denser zones correspond with the membrane surface.

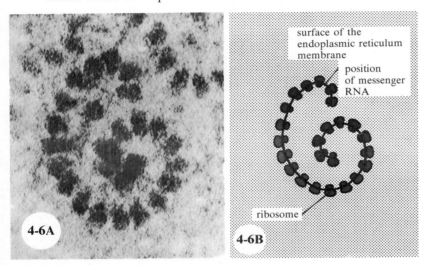

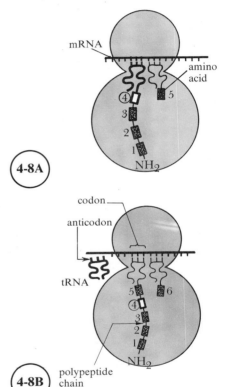

4-6A 4-6B

4-8A

4-8B polypeptide chain

4-6. Polyribosome with interpretative diagram.

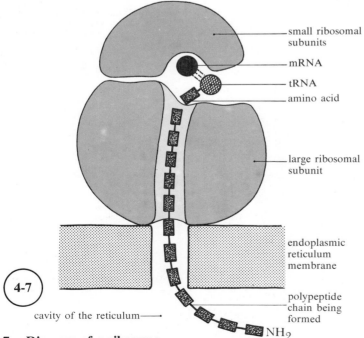

4-7

cavity of the reticulum

4-7. Diagram of a ribosome.
The subunits are shown with the probable sites for attachment of messenger RNA (mRNA) and transfer RNA (tRNA) and for the forming polypeptide chain.

4-8. Protein synthesis on a polyribosome. The polyribosome is a kind of polyvalent workbench where protein molecules are constructed. Specific genetic information is carried by mRNA in the form of a nucleotide sequence. This is read by the tRNA. Each tRNA binds and transports a given amino acid. Two complementary nucleotide triplets, the codon and the anticodon, link together during the attachment of an amino acid to the polypeptide chain. This mechanism occurs in the free polyribosomes of the cytosol as well as on polyribosomes attached to the membrane of the endoplasmic reticulum.

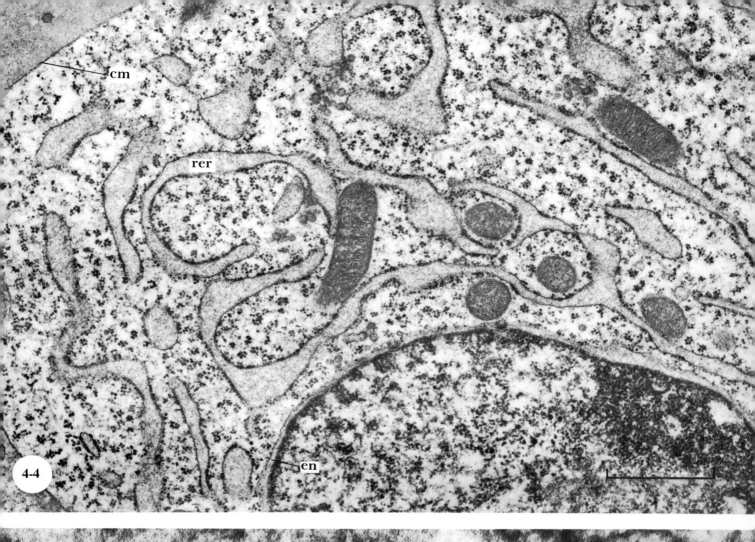

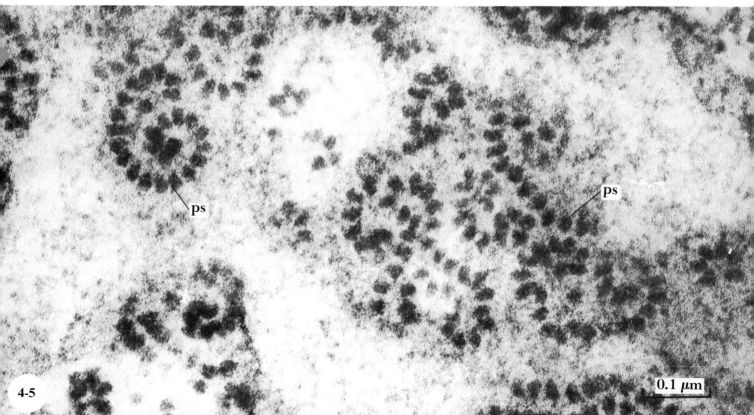

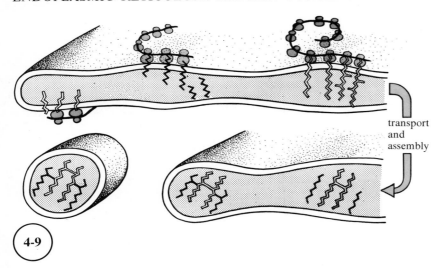

4-9

4-9. Protein transport and assembly in the endoplasmic reticulum.

In a number of cellular types, polypeptide chains synthesized by polyribosomes associated with membranes are secreted into the cavity of the reticulum. There they can associate into more complex protein structures. Some of these proteins have a primary physiologic role as antibodies, enzymes, and so on.

transport and assembly

The following three figures demonstrate the proteins in the endoplasmic reticulum having enzymatic activity.

4-10. Localization of phosphatases. ×25,000.

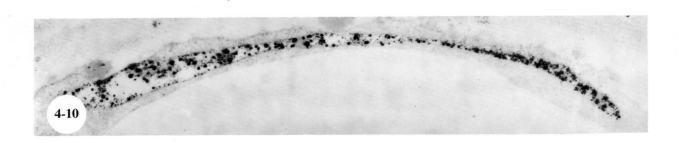

4-10

4-11. Localization of peroxidases. Cell of rat salivary gland, ×13,000. (A. Novikoff et al.)

4-12. Localization of peroxidases. Rabbit leukocyte, ×50,000. (D. Bainton and M. Farquhar.)
There is a positive reaction in the nuclear envelope (en), endoplasmic reticulum (er), and in secretory granules (sg). Note the relationship of continuity between the nuclear envelope and endoplasmic reticulum (*arrows*). Other structures do not have peroxidase activity. nu = nucleolus; ch = chromatin; m = mitochondrion; cm = cytoplasmic membrane.

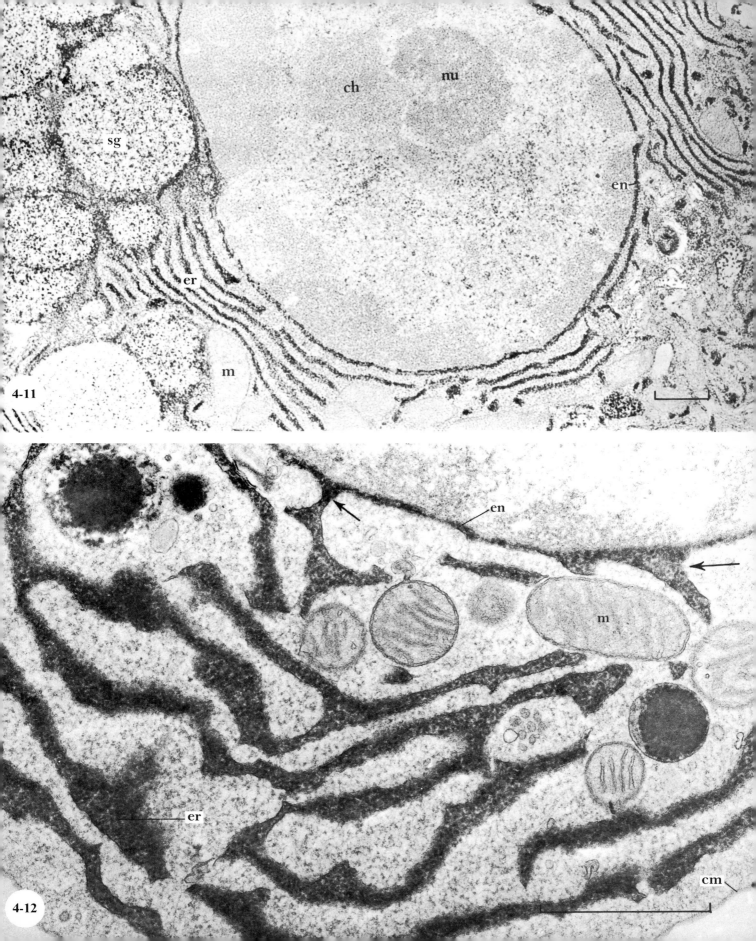

4-13. Golgi apparatus cut in different planes. Rat oocyte, ×31,000.

4-14. Two Golgi apparatuses after freeze-etching. Pea root, ×95,000. (B. Vian.)
In A there is a surface view of the saccule showing a fenestrated border and the peripheral budding of vesicles. The fracture of Golgi apparatus, B, shows the scale-like overlapping of saccules.

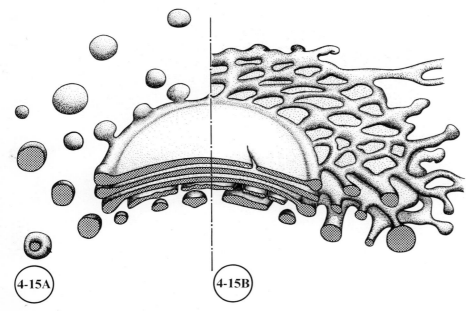

4-15A   4-15B

4-15. Spatial reconstruction of Golgi apparatus.
A. As stacked saccules and associated vesicles.
B. As fenestrated peripheral saccules.

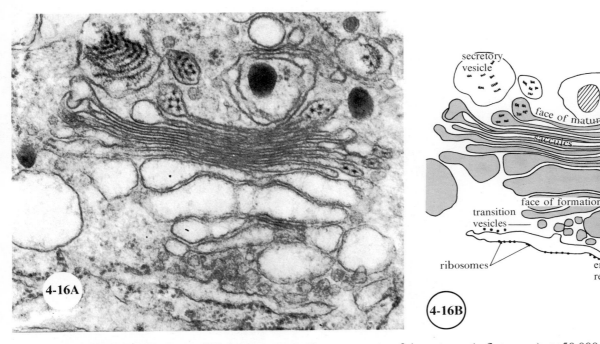

4-16A

4-16B

4-16. Progressive development of Golgi apparatus. Spermatocyte of Anapterus (a flatworm), ×50,000. (M. Silveira.) Autoradiographic, morphologic, and biochemical data together show that there is flow of secretory product from the rough endoplasmic reticulum to the secretory vesicles. One may distinguish, thus, a face of formation and a face of maturation. The photograph on the left is diagrammed on the right; the *arrow* in the diagram indicates the direction of flow of the membrane.

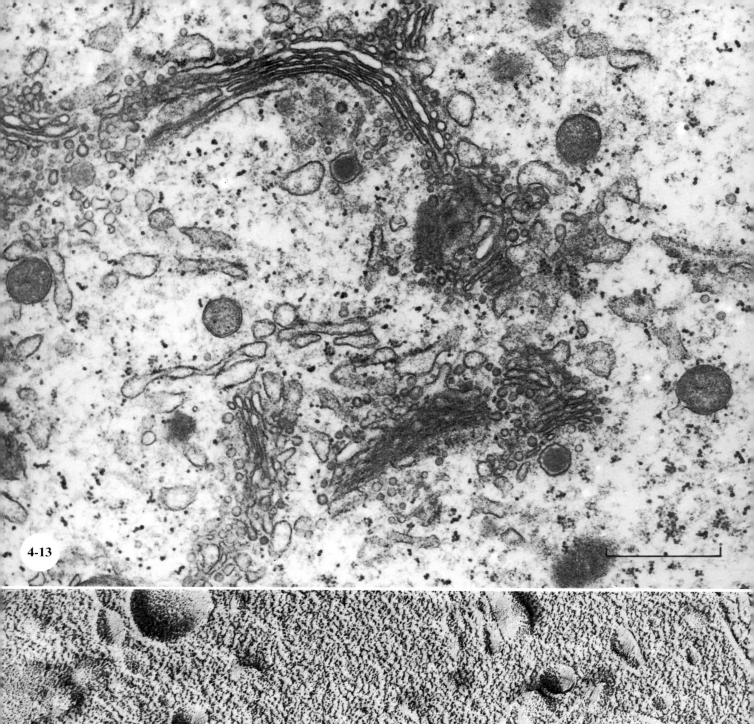

4-13

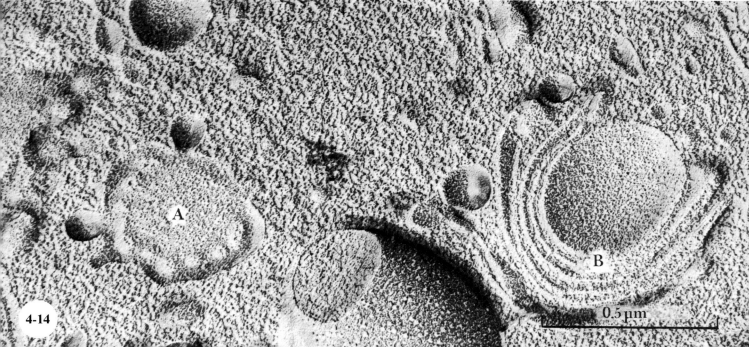

4-14

A

B

0.5 µm

GOLGI APPARATUS

Golgi bodies are a part of a complex cellular network: the Golgi apparatus. Though discovered toward the end of the last century, the existence of this network nevertheless has been discussed and even denied until recently. The existence of the Golgi apparatus has been demonstrated perfectly by very high voltage electron microscopy, which permits exploration of relatively thick cellular sections (Figs. 4-17, 4-18, and 4-19).

4-17. Ultrathin section. Mucous gland of the snail, ×15,000. (P. Favard and N. Carasso.)
Numerous Golgi bodies elaborate secretory granules (sg) of mucus, which are discharged at the apex of cells. The tissue was impregnated with osmium, which selectively contrasts the saccule of the forming or entry face in each Golgi body. Most Golgi bodies (g) are seen in transverse section. The *arrow* indicates a tangential view of the sac.

4-18 and 4-19. Thick sections (4 μm). Mucous gland of the snail, ×20,000 and ×15,000. (P. Favard and N. Carasso.) These are examined at a potential difference of 2.5 million volts, 20 to 30,000 times greater than that applied in the ordinary electron microscope. Saccules at the entry face (impregnated with osmium) establish a relationship between different Golgi bodies. Their fenestrated nature is well seen. m = mitochondrion; sg = secretory granule; g = Golgi apparatus.

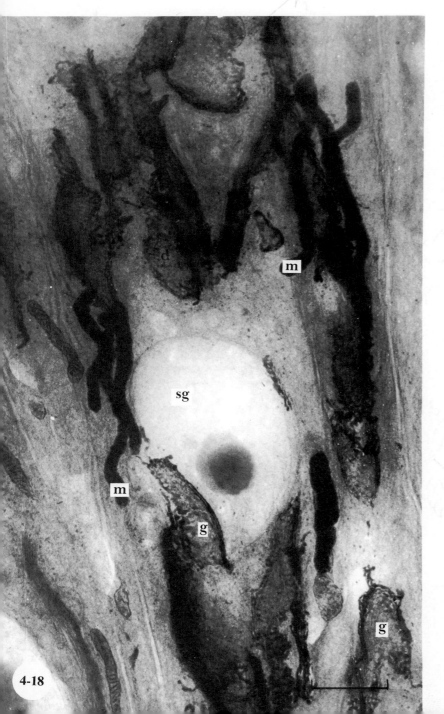

4-18

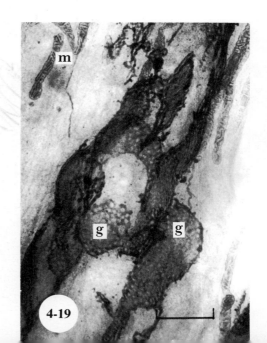

4-19

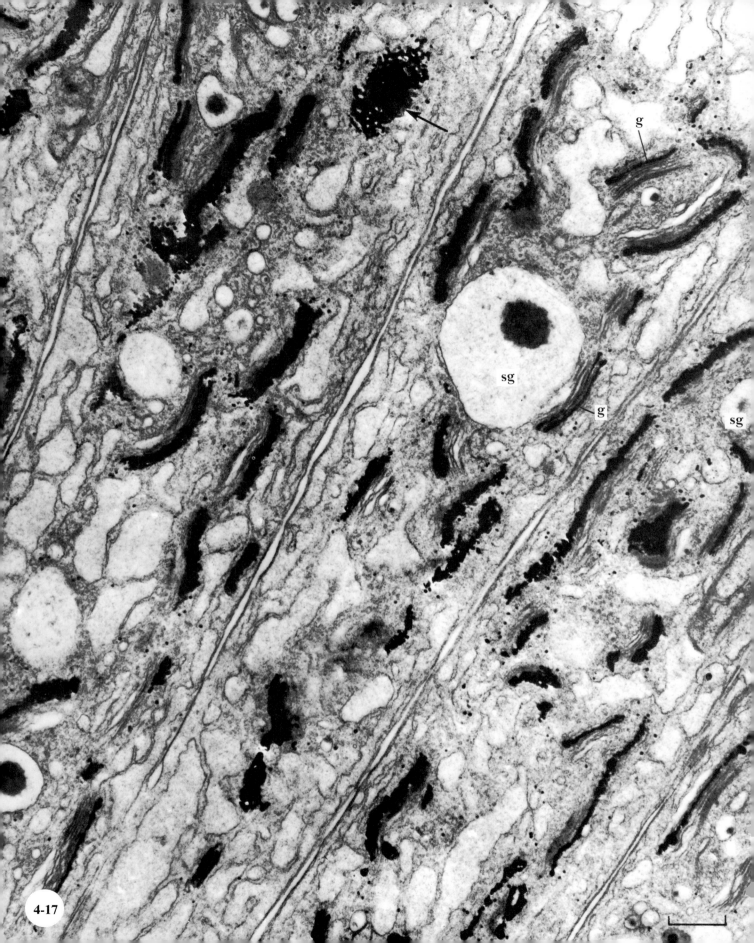

4-17

GOLGI APPARATUS: FUNCTION

The Golgi apparatus is central in the phenomenon of cell secretion. Proteins are synthesized in the rough endoplasmic reticulum and transported to this organelle via the smooth surfaced transporting vesicles budded off from the transitional elements of the rough endoplasmic reticulum. Here they are modified: polysaccharide side chains of glycoproteins are added, some proteolysis occurs, and all transported products are concentrated and eventually budded off from the Golgi apparatus as smooth membrane–bounded secretory granules. In the case of plant cells, the cellulose cell wall is synthesized in the Golgi apparatus by polymerization of monosaccharide subunits from nucleotide (activated) sugars.

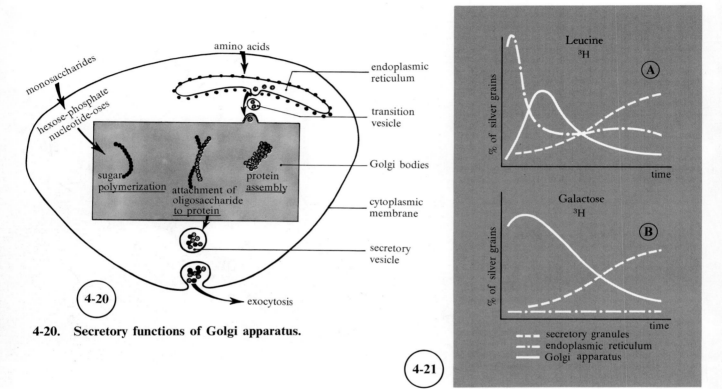

4-20. Secretory functions of Golgi apparatus.

4-21. Demonstration by radiography of the migration of products elaborated in the cell.
   A. A protein-secreting cell was incubated briefly in medium containing radioactive amino acid (^{3}H-leucine). Labeled proteins can be detected by autoradiography (% silver grains) initially in the endoplasmic reticulum. After removal of the free radioactive precursor, the labeled protein moves to the Golgi apparatus where it is modified (including addition of polysaccharide side chains as seen in **B** below) and finally packaged into secretory granules.
   B. As above, but incubated briefly in medium containing a radioactive sugar (^{3}H-galactose). Incorporation is initially in the Golgi apparatus where it is subsequently found in secretion granules, again detected by autoradiography.

Figures 4-22 and 4-23 demonstrate the cytochemical detection of polysaccharides and enzymes within a Golgi body. The tissue was germinal male cell during acrosome elaboration (an acrosome is a secretory granule situated specifically in the head of a spermatozoon).

4-22. Polysaccharides. Batrachian testicle, $\times$69,000. (D. Sandoz.)
   Polysaccharides are made in the saccules and are concentrated and transported in the Golgi vesicles (g), which fuse to form the acrosome (ac). d = dictyosome.

4-23. A phosphatase enzyme. Mouse testicle, $\times$60,000. (D. Sandoz.)
   The activity of the phosphatase enzyme is selectively localized to the exit face of a Golgi body (g).

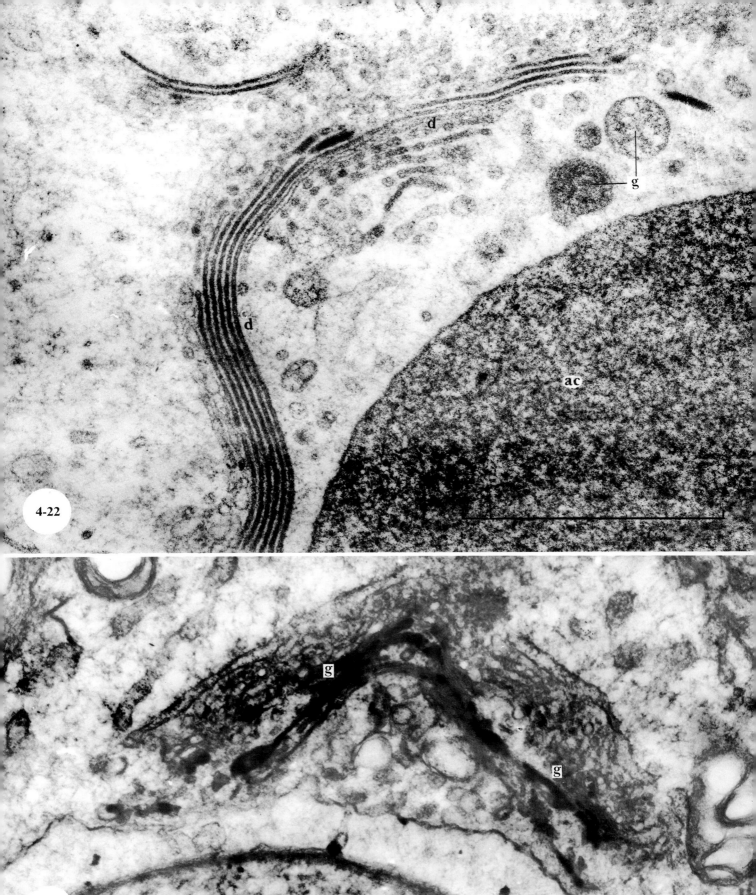

Emptying of the zymogen granule begins when the granule, triggered by hormonal stimuli, approaches the membrane at cell apex (1). The two membranes first touch (2) and then merge (3); finally, the merged area opens (4) to allow ejection of the granule's protein contents. (Reprinted with permission from J. D. Jamieson, Membranes and Secretion, *Hosp. Pract.* 8:12, 1973, and G. Weissmann and R. Claiborne, eds., *Cell Membranes: Biochemistry, Cell Biology, and Pathology,* New York: HP Publishing Co., Inc., 1975.)

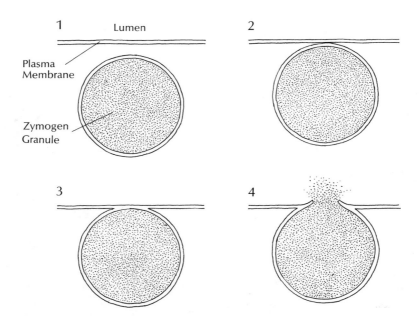

4-24. **Tracking of protein through cell compartments by means of pulse chase autoradiography.** (Reprinted with permission from J. D. Jamieson, Membranes and Secretion, *Hosp. Pract.* 8:12, 1973, and G. Weissmann and R. Claiborne, eds., *Cell Membranes: Biochemistry, Cell Biology, and Pathology,* New York: HP Publishing Co., Inc., 1975.)

A. Immediately following labeling (3 minutes), label (*arrows*) appears in the rough endoplasmic reticulum (RER), showing presence of newly synthesized proteins.

B. By 7 minutes with cells in unlabeled medium, pulse of labeled protein has moved into vesicles and cisternae of the Golgi apparatus.

C. At 37 minutes most of the label is in the condensing vesicle (CV).

D. At 80 minutes nearly all the label is in the darker zymogen granules (ZG) or has passed into the lumen (L).

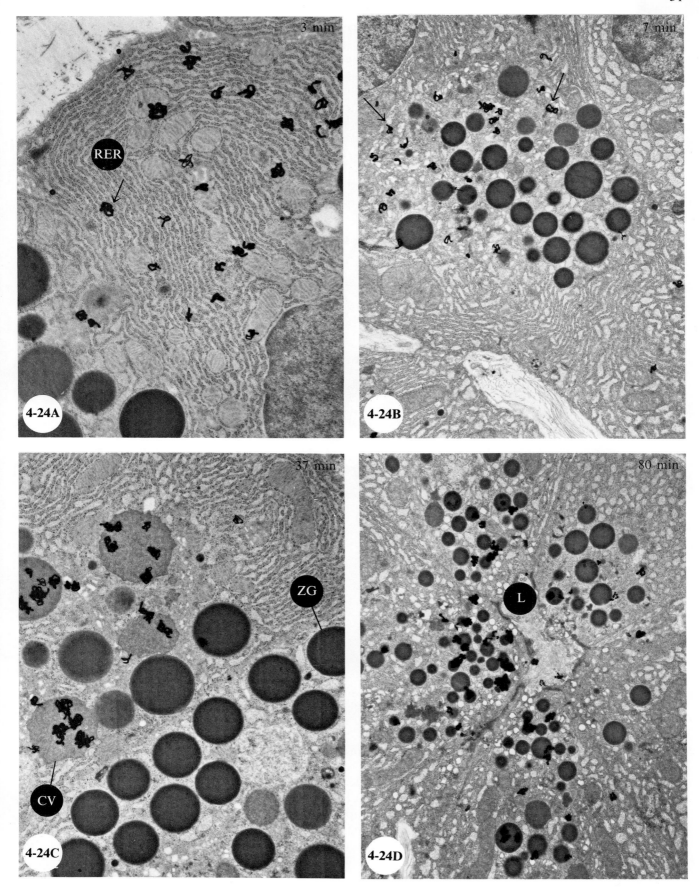

3 min

RER

4-24A

7 min

4-24B

37 min

ZG

CV

4-24C

80 min

L

4-24D

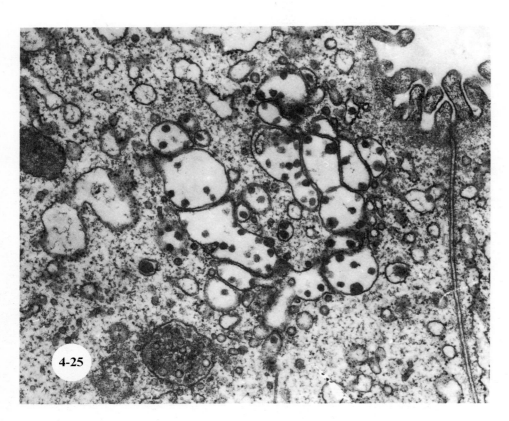

4-25

4-25. Concentrations of lipo-proteins in Golgi apparatus. Frog liver, ×40,000.

4-26. Glycoproteins and protein granules of Golgi apparatus origin in an endocrine gland. Toad hypophysis, ×7500. (F. Mira Moser.)

The hypophysis contains several types of cells characterized by their secretory granules, which are varied in form and in nature. These granules contain hormones. At the center of the picture, the three cells containing small glycoprotein granules (about 2000 Å) secrete thyrotropic hormone, which acts upon the thyroid gland. On either side of this group of cells, two cells are seen containing larger granules, glycoprotein or protein in nature, which secrete other hypophyseal hormones. The contents of these vesicles are excreted and hormones enter the circulation.

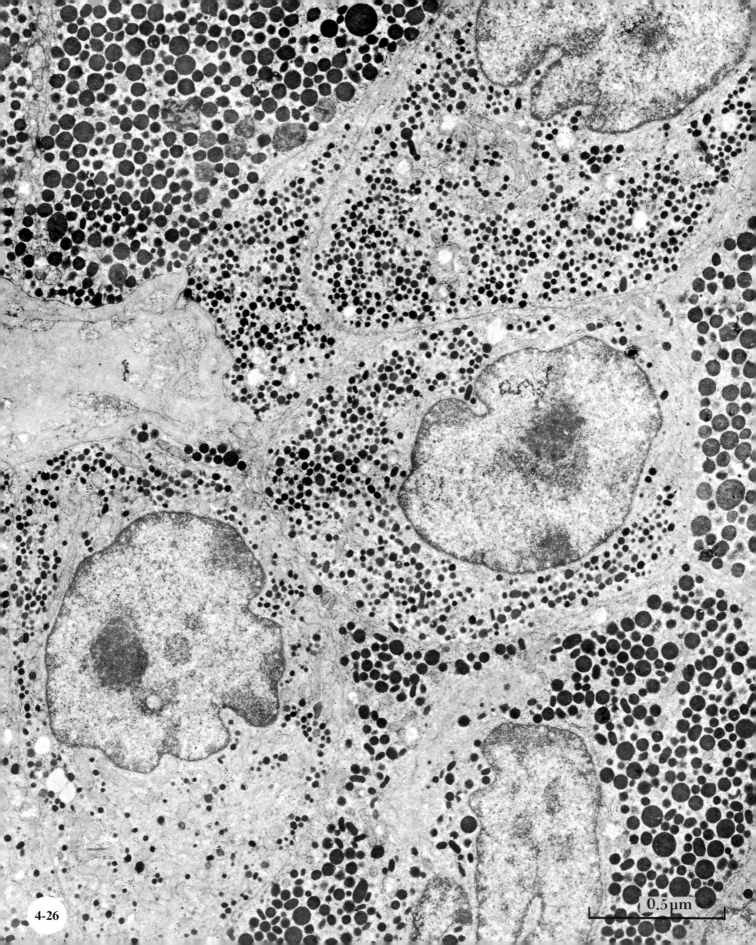

4-26

0.5 μm

LYSOSOMES

Lysosomes are compartments of the internal membrane system and have a lytic function. They take part in the digestion of exogenous substances and in lysis of parts of the cell itself (autophagia). They are so rich in hydrolytic enzymes that they are virtually capable of destroying all of the cellular constituents.

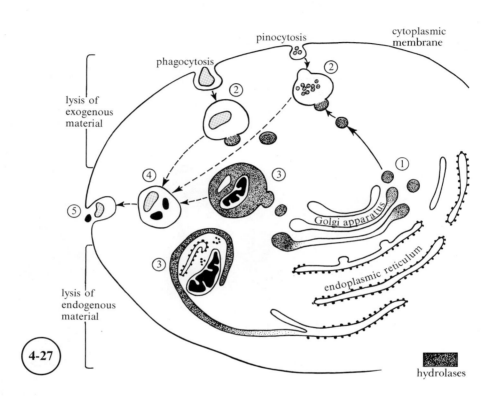

4-27. Lytic system of the cell.
1. Formation of primary lysosomes (Golgi bodies).
2. Lysis of exogenous material: Digestive vacuoles or secondary lysosomes form by fusion of primary lysosomes bearing hydrolases with the vesicles of pinocytosis and phagocytosis.
3. Lysis of endogenous material: Portions of the cytoplasm are sequestrated by a diverticulum of the smooth endoplasmic reticulum containing hydrolases. An autophagic vacuole forms, then primary lysosomes pour their contents into it.
4. Residual bodies are formed by the accumulation of undigested products.
5. Exocytosis eventually eliminates the residues into the extracellular medium.

4-28. Lysosomes in a phagocytic cell. Sponge archeocyte, ×30,000. (L. de Vos.)
The numerous lysosomes present in the cytoplasm are digesting cellular fragments of exogenous origin (residues of neighboring degenerate cells). These are secondary lysosomes (ly). n = nucleus; m = mitochondrion; g = Golgi apparatus; l = lipid droplets; cm = cytoplasmic membrane; rer = rough endoplasmic reticulum.

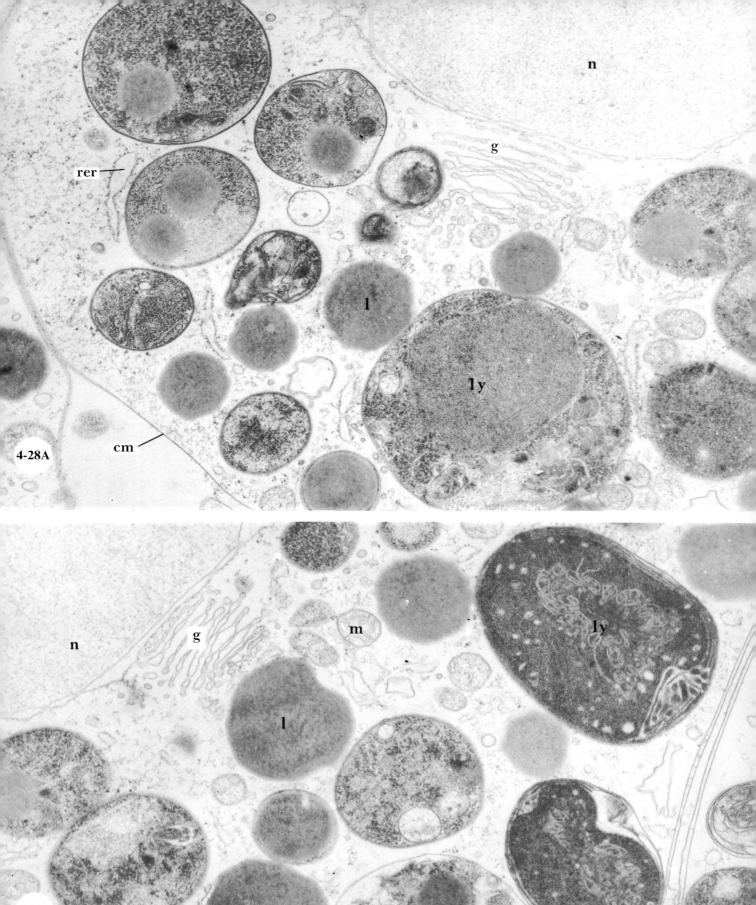

PEROXISOMES

Other compartments of the internal membrane system can be identified. Peroxisomes are the compartments in which oxidation-reduction enzymes are segregated. Sometimes one can detect hydrolases and oxidases simultaneously, but a difference tends to be established, and typical peroxisomes essentially contain enzymes which metabolize peroxides.

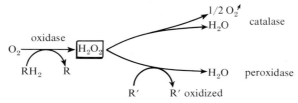

Peroxisomes are frequently associated with the other organelles which take part in oxidation-reduction reactions: mitochondria and chloroplasts. In plants, peroxisomes may be the site of photorespiration, with metabolites furnished by chloroplasts producing carbon dioxide. Figures 4-29 and 4-30 show the concentration of peroxidases and acid phosphatases in cytoplasmic granules.

4-29. **Peroxidases.** Young eosinophil leukocyte of rabbit, $\times 15,000$. (D. Bainton and M. Farquhar.)

At this stage of development, peroxidases are localized in the endoplasmic reticulum (er), in the Golgi apparatus (g), and in the cytoplasmic granules. In the more mature cells, enzymatic activity becomes concentrated in the granules and disappears from the reticulum and the Golgi apparatus.

4-30. **Acid phosphatases.** Young eosinophil leukocyte of rabbit, $\times 24,000$. (D. Bainton and M. Farquhar.)

Granules are made up of a matrix and a crystal. Only the matrix reacts positively.

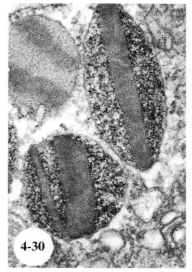

4-31 and 4-32. Detection of catalases in peroxisomes. Tobacco leaf, $\times 50,000$ and $\times 38,000$. (S. Frederick and E. Newcomb.)

These pictures show the three organelles implicated in oxidation-reduction phenomena: peroxisomes (px), mitochondria (m), and chloroplasts (cpl). Catalase alone is present in the peroxisome, giving the heightened contrast to this organelle. Note the tight association between the peroxisome and the chloroplast, which probably indicates the transfer of metabolites from the chloroplast to the peroxisome.

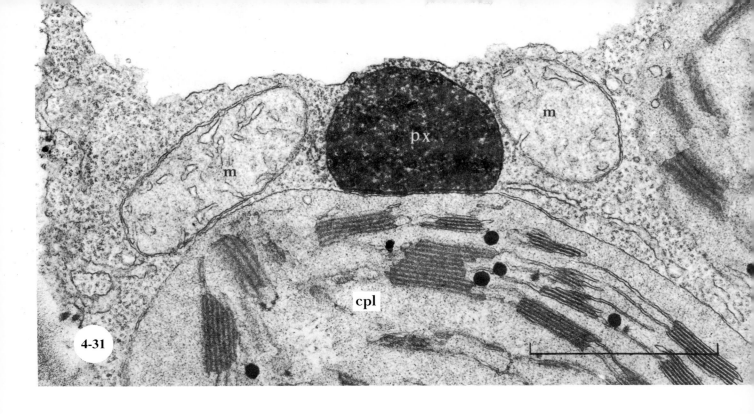

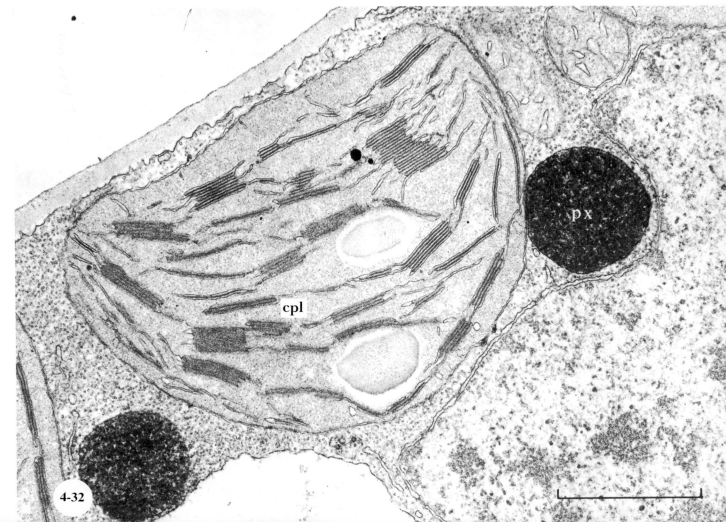

5. SEMIAUTONOMOUS ORGANELLES

Mitochondria and plastids differ from other cellular organelles in several respects. (1) They have their own genetic material (DNA), which assures them a certain autonomy relative to the nucleus and enables them to synthesize some of their proteins. This material is transmitted when these organelles divide, so they have genetic continuity. (2) They are morphologically well defined by two limiting membranes, the inner of which is invaginated into cristae and lamellae and is highly differentiated. (3) Their physiology is oriented toward ATP production (Fig. 5-1). This molecule of low molecular weight is a bearer of energy which acts in numerous steps of cellular metabolism according to the reversible reaction:

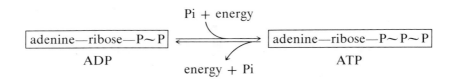

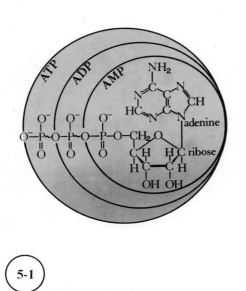

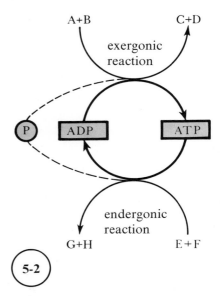

5-1. **ATP structure.**
The symbol $\sim$ designates an energy-rich bond.

5-2. **Energy transfer** through ATP as the intermediary between reactions which produce energy (exergonic) and reactions which consume energy (endergonic). The latter are highly varied and include biosynthesis and muscular contractions.

MITOCHONDRIA

Figures 5-3 and 5-4 show general views of a chondriome (an assembly of mitochondria in a cell).

5-3. **Chondriome.** Frog liver cells, $\times 10,000$.
n = nucleus; m = mitochondrion; bc = bile canaliculus.

5-4. **Chondriome.** Hamster oocyte, $\times 40,000$.
The two limiting membranes are distinguishable. icr = internal cristae; mx = mitochondrial matrix. Here and there the mitochondria are closely apposed to one another. ic = intermitochondrial cement.

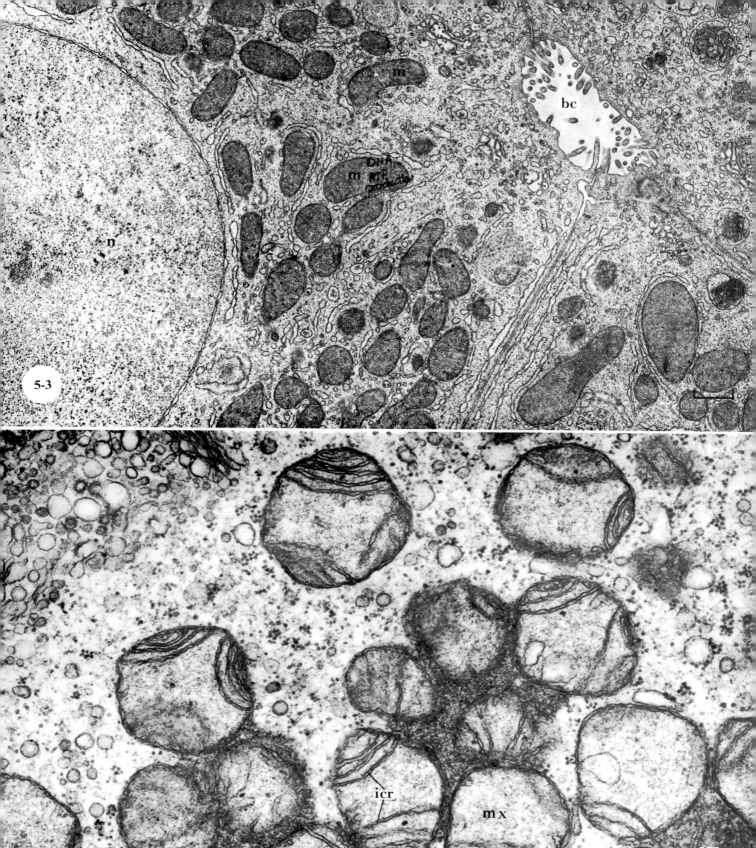

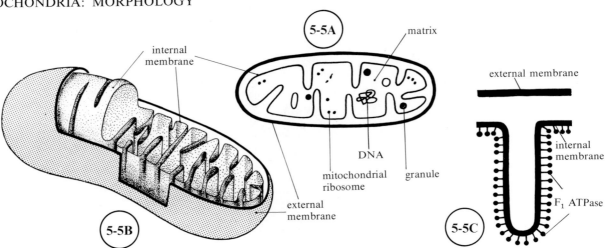

5-5. Organization of a mitochondrion.
(A) Section. (B) Schematic three-dimensional representation. (C) Detail of a crista, showing the pedunculated particles (ATPosomes) of the internal membrane.

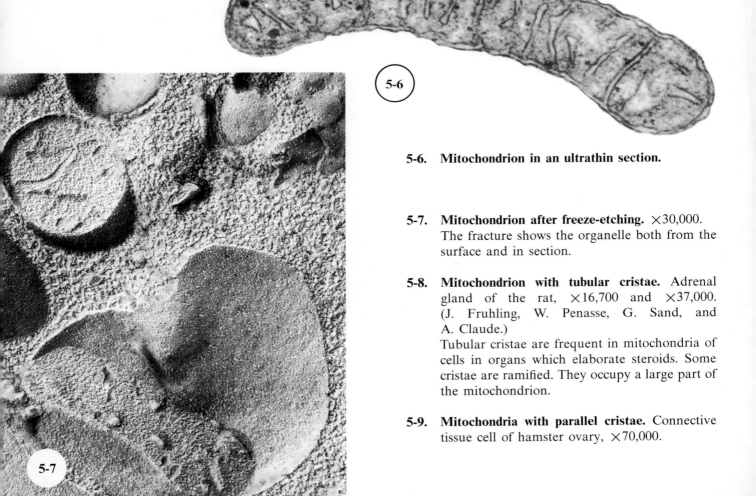

5-6. Mitochondrion in an ultrathin section.

5-7. Mitochondrion after freeze-etching. ×30,000. The fracture shows the organelle both from the surface and in section.

5-8. Mitochondrion with tubular cristae. Adrenal gland of the rat, ×16,700 and ×37,000. (J. Fruhling, W. Penasse, G. Sand, and A. Claude.)
Tubular cristae are frequent in mitochondria of cells in organs which elaborate steroids. Some cristae are ramified. They occupy a large part of the mitochondrion.

5-9. Mitochondria with parallel cristae. Connective tissue cell of hamster ovary, ×70,000.

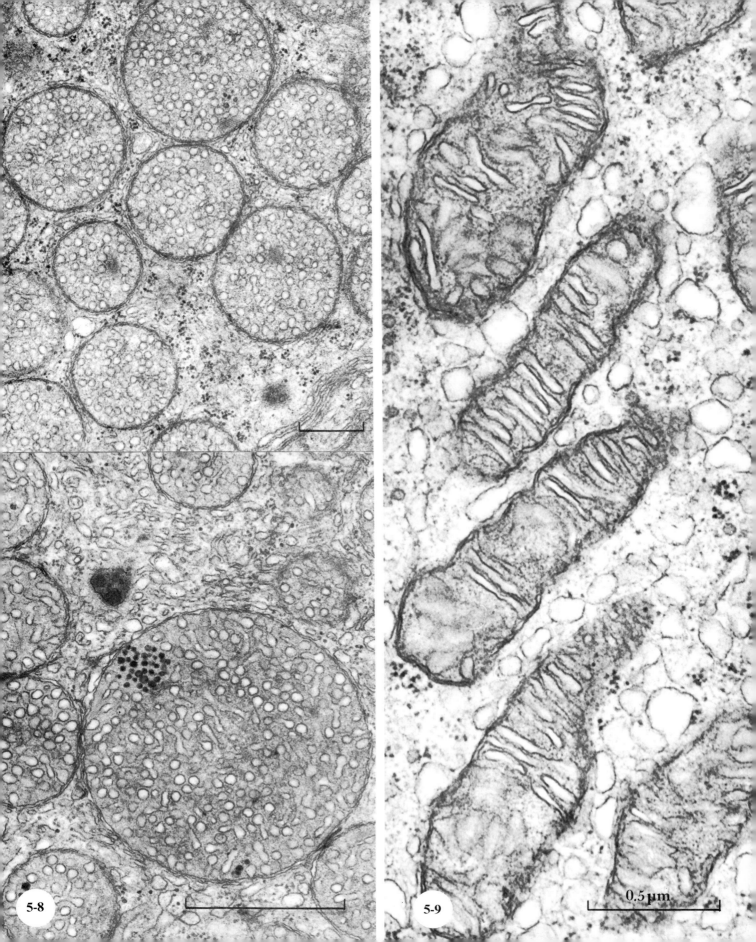

5-8

5-9

0.5 µm

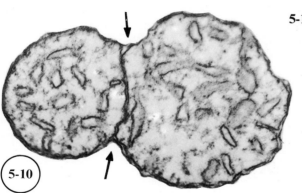

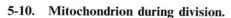

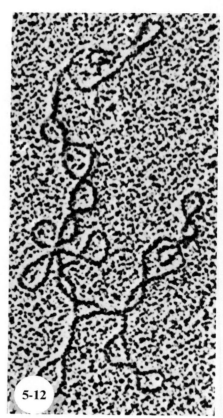

5-11. Mitochondrial DNA in situ. ×74,000.
In the matrix, a clear space contains coils of DNA fibrils 20 Å in diameter (*arrow*). There are mitochondrial ribosomes in the encircled areas. rb = cytoplasmic ribosomes.

5-10

5-10. Mitochondrion during division.

Figures 5-12 and 5-13 show mitochondrial DNA isolated from rat liver tissue. First the mitochondria were isolated; then they were burst by osmotic shock. Collected DNA molecules were subjected to a treatment which increased their thickness. After shadowing, they were examined.

5-12. Isolated mitochondrial DNA. Rat liver, ×50,000. (M. Guerineau.)
In the native state, DNA molecules remain in coils.

5-13. Isolated mitochondrial DNA. Rat liver, ×50,000. (M. Guerineau.)
One can open the molecule and extend it by breaking one of the two threads of the double helix by irradiation (x-rays). This molecule is circular. It measures 5 μm in animal cells.

5-12

5-14. Mitochondrial polyribosomes. Mitochondria isolated from yeast, ×36,000. (P. Vignais, B. Stevens, J. Huet, and J. André.)
The mitochondrial ribosomes are associated in polyribosomes and arranged along the internal membranes. The polyribosomes serve in the synthesis of a portion of mitochondrial proteins: they translate the genetic information carried by the mitochondrial DNA.

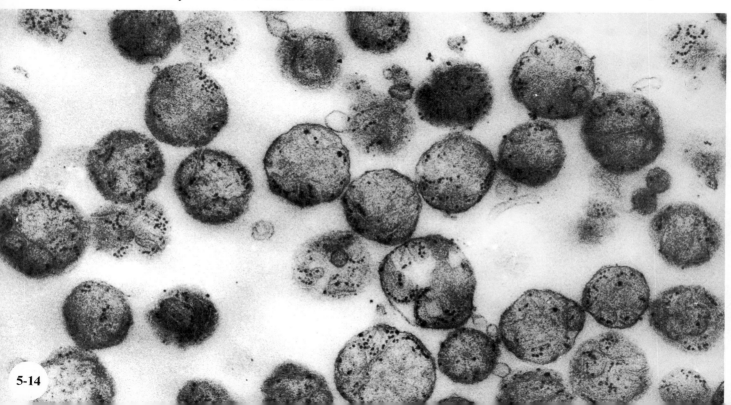

5-14

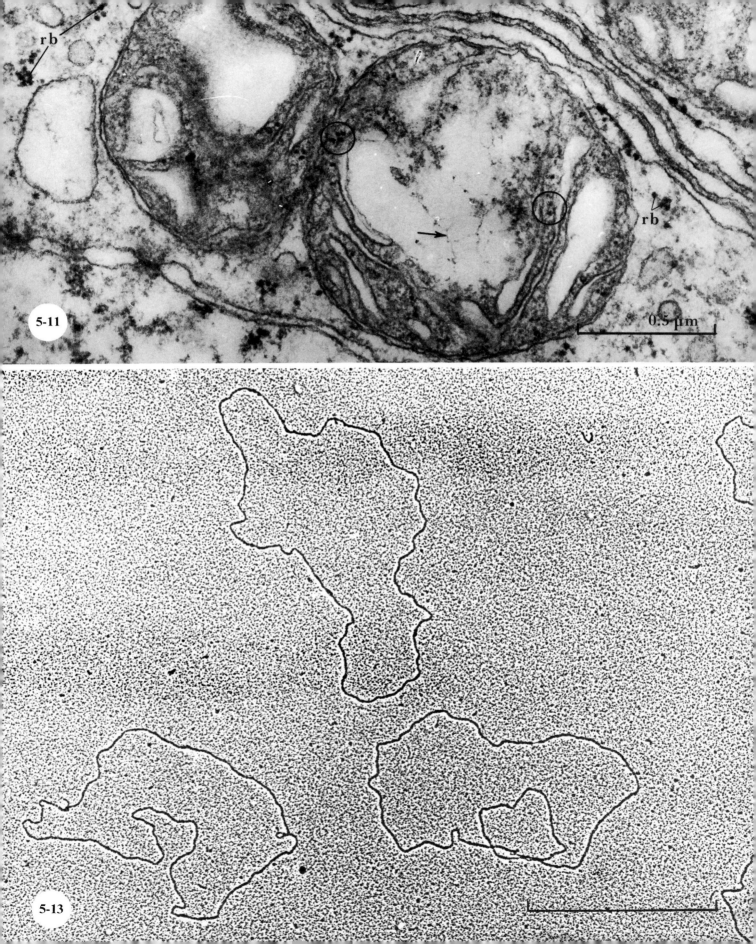

5-11

rb

rb

0.5 μm

5-13

MITOCHONDRIA: FUNCTION

Mitochondria are the site of cellular respiration, during which ATP is formed by oxidative phosphorylation. They can be considered to be the energy centers of the cell.

5-15. Detection of oxidation-reduction activity in the mitochondrion. Rat liver, ×36,000. (A. Seligman et al.) The enzymatic activity revealed is that of cytochrome oxidase, which acts in the transport of electrons. Reaction products accumulate in the cristae and in the intermembranous space (*arrows*). er = endoplasmic reticulum.

5-16. Demonstration of particles of the internal mitochondrial membrane. Soybean after negative staining, ×250,000. Internal membranes were isolated for this study. Seen in profile are numerous pedunculated particles 100 Å in diameter (*arrows*). They are distributed along the entire internal surface of the membrane. These particles were called *oxysomes* at the time of their discovery because they were attributed a role in oxidation-reduction reactions of the respiratory chain. Biochemical analyses indicate that, in fact, they are the site of ATP production.

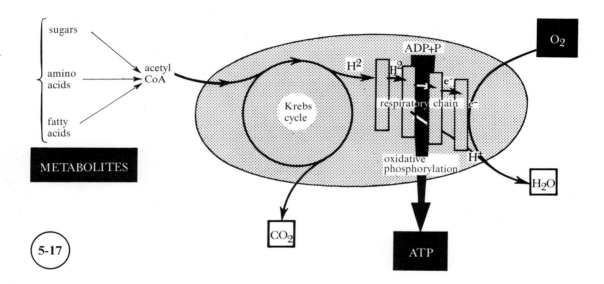

5-17.

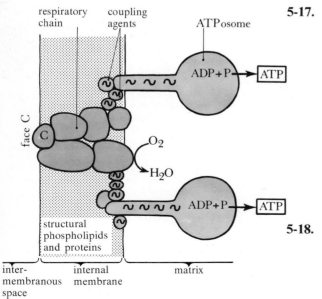

5-18.

5-17. Cellular respiration in the mitochondrion.
Acetylcoenzyme A (acetyl CoA) coming from cytoplasmic metabolites is integrated into a cycle of reactions (Krebs cycle), during which it is decarboxylated (carbon dioxide is removed) and oxidized by loss of hydrogen. The hydrogen enters an oxidation-reduction chain (respiratory chain) that is formed by specific carriers of hydrogen or of electrons: NAD, NAD + 1, cytochromes. Energy liberated during oxidation-reduction reactions permits ADP phosphorylation (oxidative phosphorylation). Hydrogen combines with oxygen to form water. The enzymes of the Krebs cycle have been identified in the matrix. Transporters of the respiratory chain and phosphorylation enzymes are associated at the internal membrane.

5-18. Organization of the internal membrane.
The coupling of oxidation-reduction and phosphorylation reactions implies a hierarchical grouping of enzymes (multienzymatic complexes) and coupling agents. The distribution of these groupings can be defined by controlled fragmentation methods. A map of the internal membrane is thus outlined. C = cytochrome C.

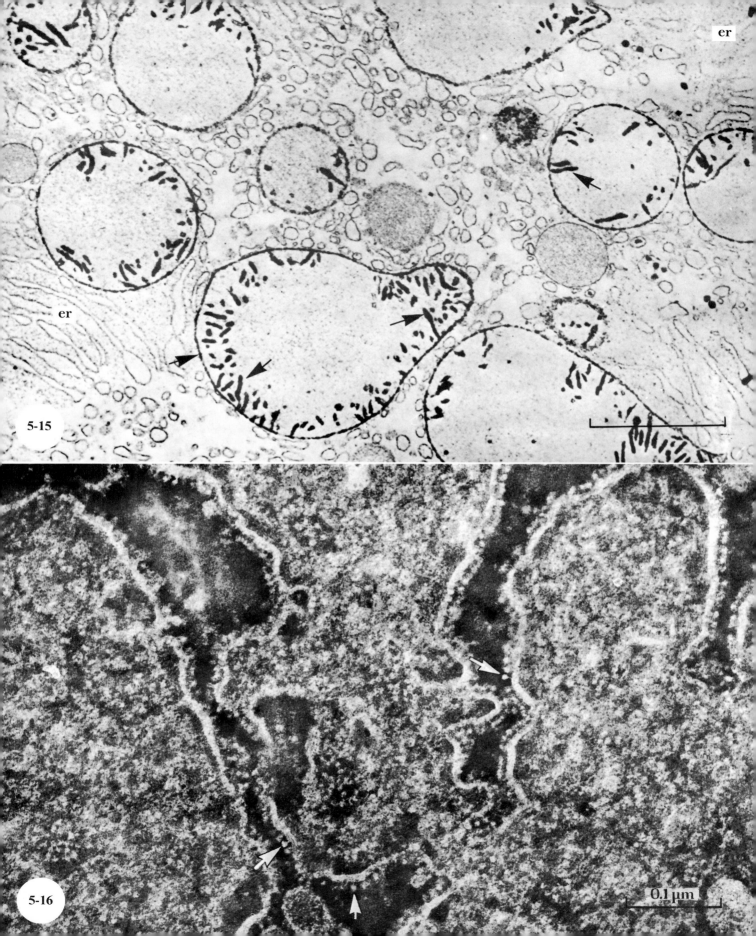

er

er

5-15

5-16

0.1 μm

MITOCHONDRIA: CONCENTRATION OF SUBSTANCES

Quite apart from their essential role in cellular respiration, mitochondria play a role in synthesis. They also have the property of concentrating such diverse substances as cations (calcium ions), glycoproteins, proteins, glycogen, and lipids.

5-19. **Accumulation of lipoprotein.** Monkey oocyte, ×50,000. Besides typical mitochondria seen here, some are modified and are able to concentrate lipoprotein granules.

5-20. **Production of protein crystals by mitochondria.** Pea root, ×100,000. (L. Leak.)

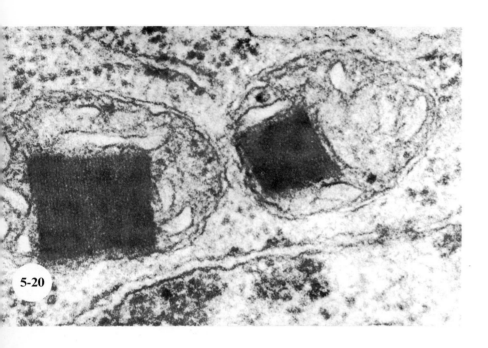

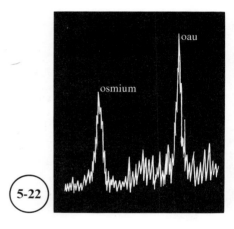

Figures 5-21 and 5-22 show the experimental concentration of a metal in a mitochondrion.

5-21. **After injecting gold salts** (which are used in the treatment of rheumatism), giant mitochondria containing a dense precipitate are seen in kidney cells. ×45,000. (P. Galle.)

5-22. **When a microprobe beam** is directed on one of these inclusions, they emit x-rays. Analysis of the spectrum indicates that it is a concentration of the injected metal (the osmium peak is due to the fixative which is bound to the tissue). (P. Galle.)

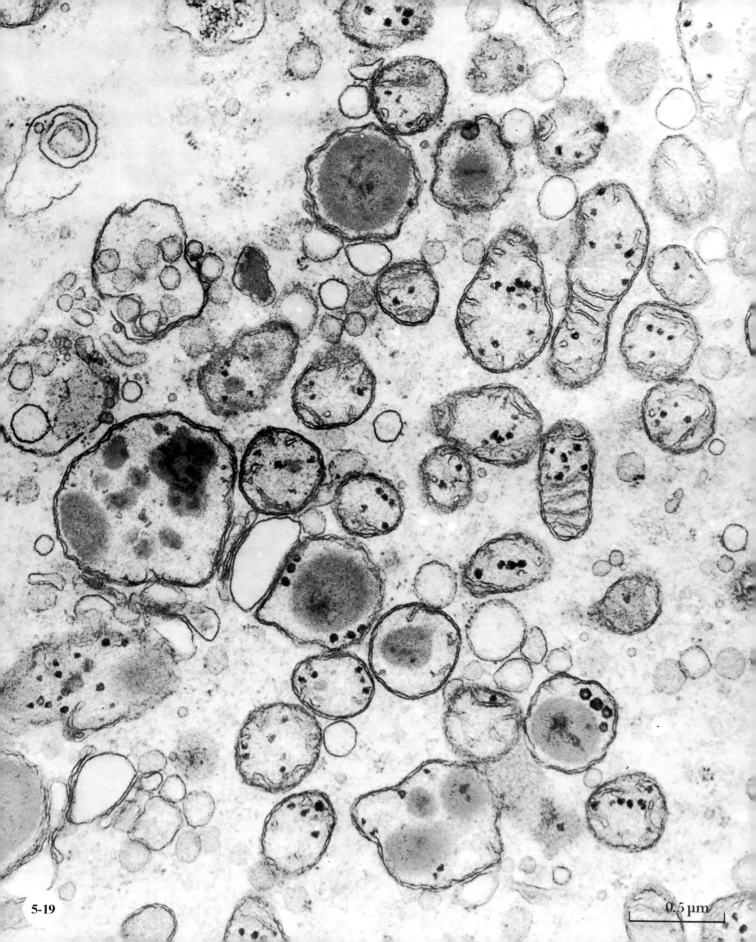

5-19

0.5 µm

MITOCHONDRIA: FUNCTIONAL ASSOCIATIONS

The grouping of certain organelles in cells suggests the existence of functional cooperation. Mitochondria furnish numerous examples. Through these associations, they furnish the necessary energy to effect a particular cellular function such as muscular contraction, flagella movement, or absorption.

5-23. **Mitochondrion association in cytoplasmic membrane.** Absorptive epithelium of an insect, ×40,000. (P. Cassier and M. Fain-Morel.)
The cytoplasmic membrane (cm) forms profound invaginations, within which are lodged mitochondria (m). This occurs in cells which are subject to intense ionic exchanges.

Figures 5-24 and 5-25 show mitochondrion association in a myofibril from skeletal muscle of a guinea pig.

5-24. **Myofibril, transverse section.** ×16,000. (M. Fardeau.) my = myofibril; m = mitochondrion; cm = cytoplasmic membrane.

5-25. **Myofibril, longitudinal section.** ×12,500. (M. Fardeau.)
The picture shows two types of skeletal fibers which are distinguishable by their differing abundance of mitochondria. my = myofibril; m = mitochondrion.

5-26. **Mitochondrion association in a flagellum.** Hamster spermatozoon, ×13,000.
Mitochondria form a sleeve around the base of a locomotor flagellum.

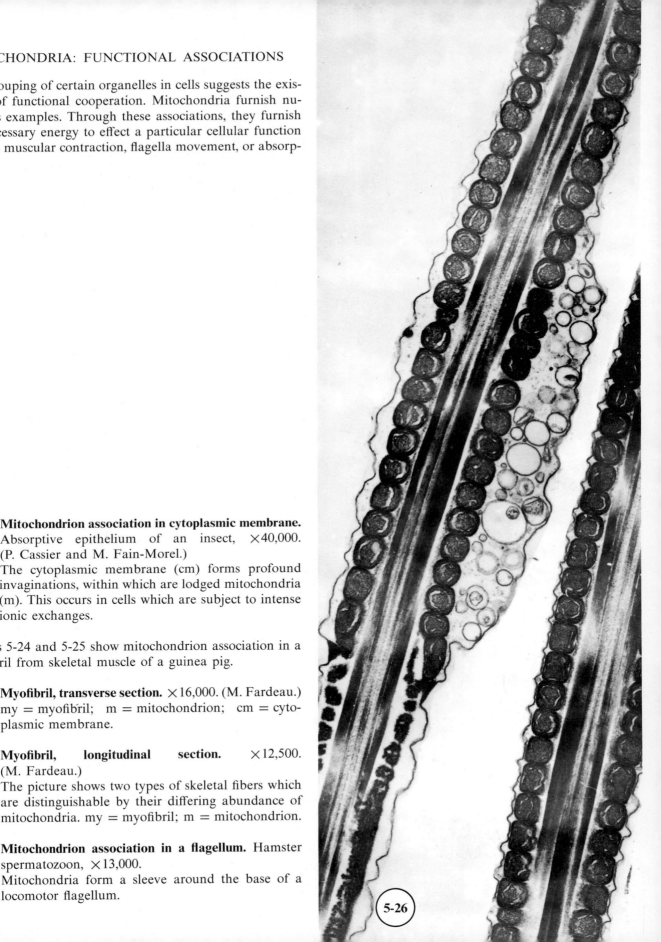

5-26

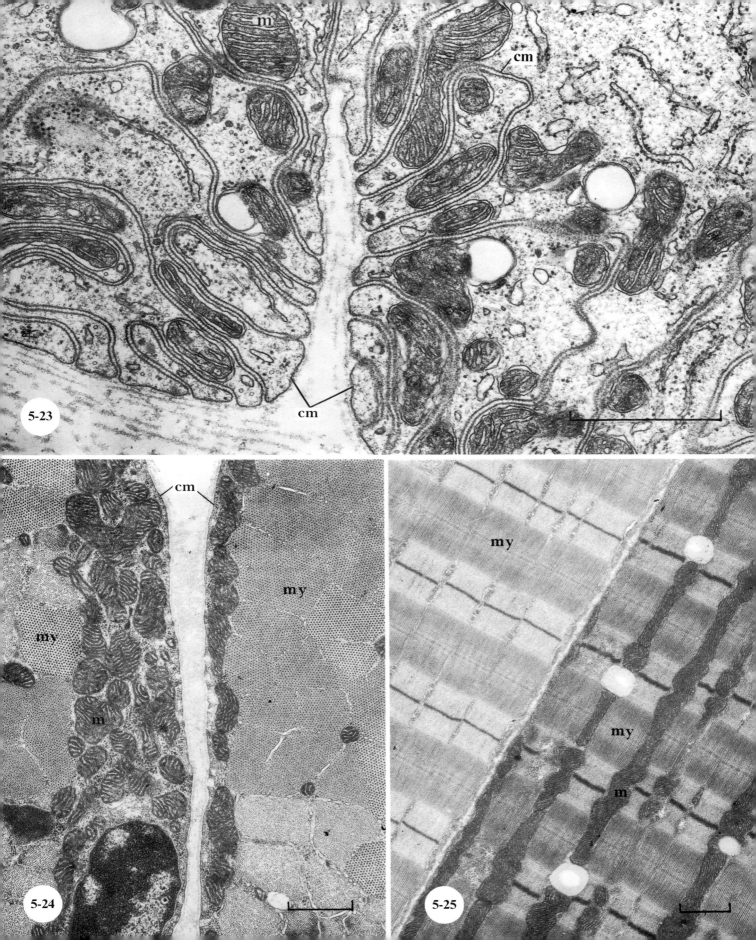

PLASTIDS: MORPHOLOGY

These organelles are present only in plant cells. The most characteristic are chloroplasts, which contain chlorophyll in a well-developed membrane system called the thylakoids. Chloroplasts are the site of photosynthesis.

5-27. Ultrathin section of chloroplast. Tobacco leaf, ×72,000. (Y. Lemoine.)

Two peripheral membranes are seen (*arrows*). s = stroma, which contains plastid ribosomes; th = thylakoids, which occur either free within the stroma of the plastid or stacked in a granum (gr); a = starch-bearing (amylaceous) vesicle; l = lipid globule.

5-28. Isolated granum, face view. Spinach leaf, ×60,000.

Stacked thylakoids are shown by negative staining. Note the clear particles on the membranes; some correspond to systems of enzymes.

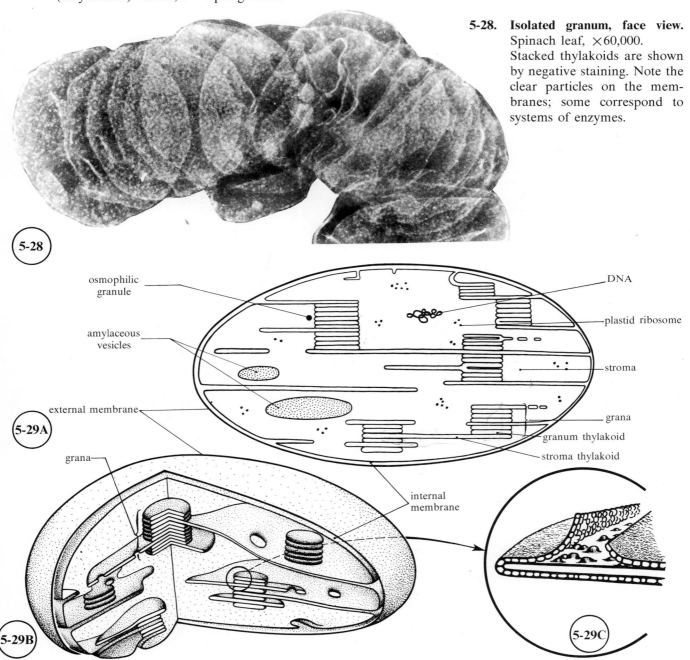

5-29. Organization of a chloroplast.
    A. Section.
    B. Schematic representation in three dimensions.
    C. Membrane of a granum thylakoid, showing its particles.

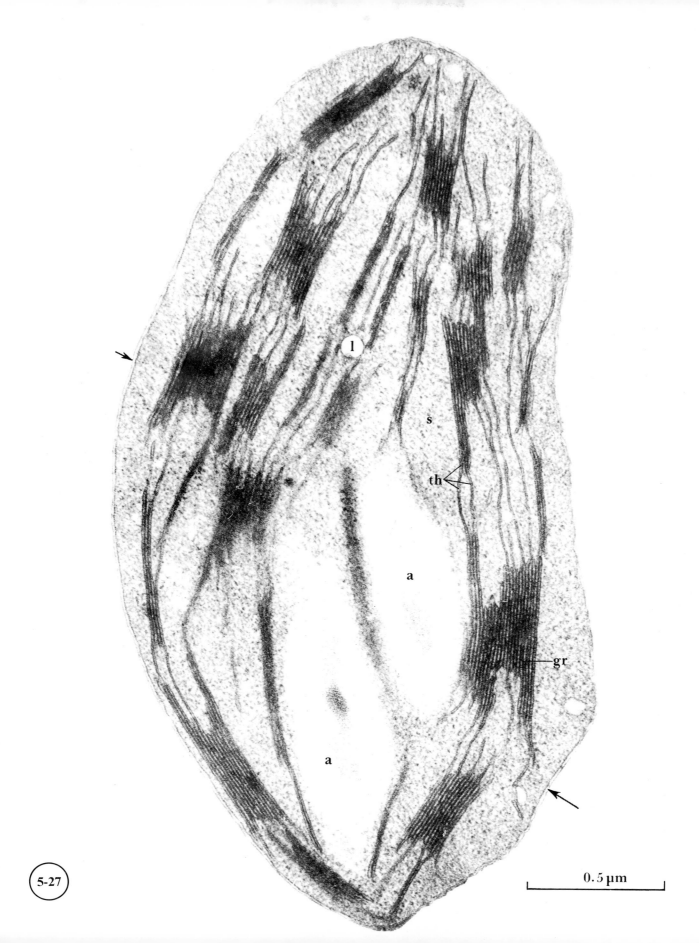

5-27   0.5 μm

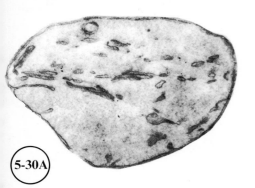

5-30A

Plastid differentiation is effected from the proplastids present in meristem cells (undifferentiated embryonic tissue). These proplastids have a simple structure, little different from that of mitochondria. Beginning with invaginations of the internal membrane, extension of thylakoids into chloroplasts can be followed during the course of cellular differentiation. It takes place in the presence of light. In the dark, either this membrane does not develop, or it regresses.

Chloroplasts divide by transverse constriction, and the genetic material, DNA, is divided between the two daughter plastids.

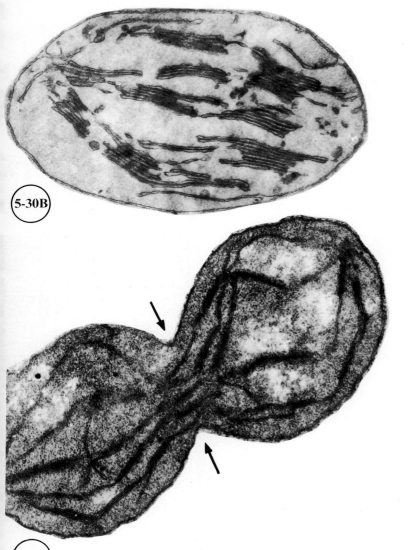

5-30B

5-30C

5-30. **Development of structure (A, B) and division of the plastids (C).** Elder leaf during growth.

Note the progressive development of the internal lamellar system (thylakoids) and the numerous invaginations of the internal membrane.

5-31. **Chloroplast DNA in situ.** Pea leaf, ×90,000.

A large number of plastid ribosomes are contained in the stroma except in the regions occupied by DNA (*clear areas; arrows*).

5-32. **DNA of an isolated chloroplast.** ×40,000. (T. Bisalputra and H. Burton.)

Note the attachment of DNA to a fragment of the plastid membrane (*arrow*).

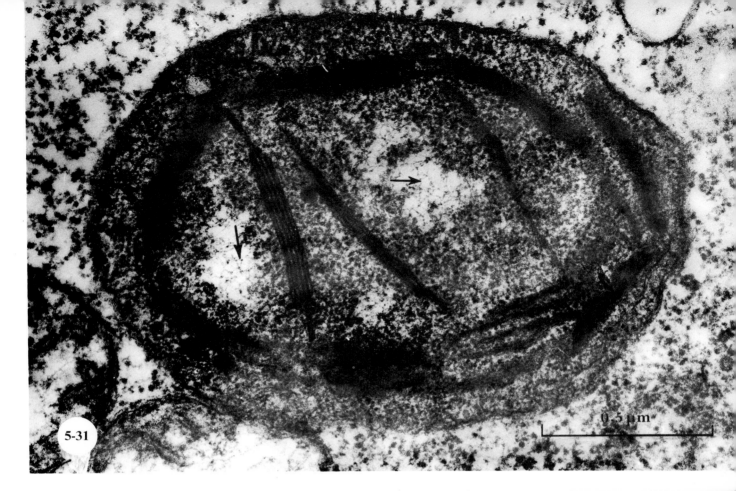

0.5 μm

5-31

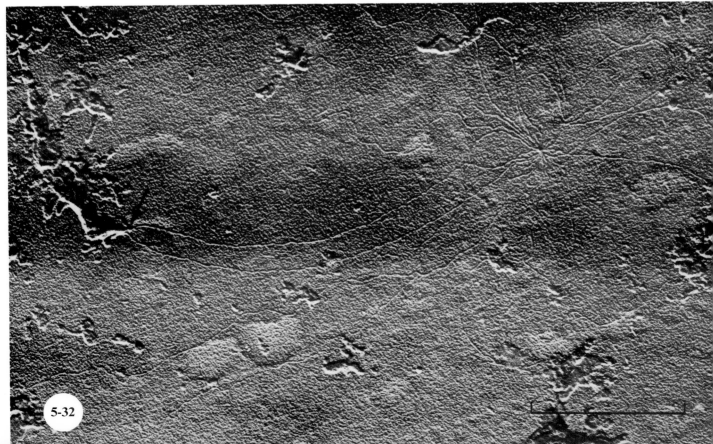

5-32

54

PLASTIDS: FUNCTION

Plastids make photosynthesis possible through utilization of light energy. Photosynthesis is a series of coordinated processes which lead to the synthesis of organic substances from water and carbon dioxide.

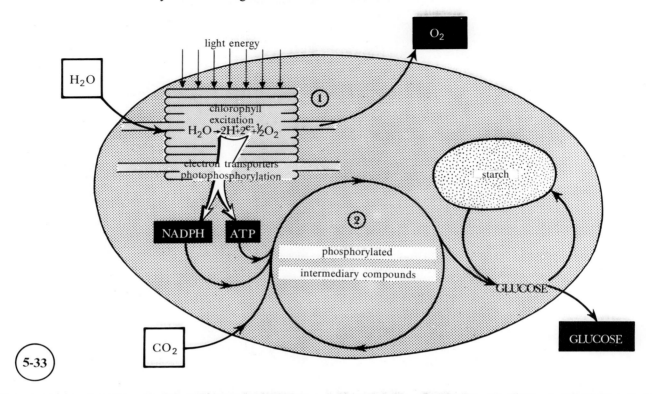

5-33

5-33. Diagrammatic summary of the process of photosynthesis.

Two phases can be distinguished: First, the light phase (1), during which there is conversion of energy into chemical energy and photolysis of water occurs. This phase results in the production of ATP and NADPH (photophosphorylation). This takes place in the thylakoids where the photosystems which capture light energy are located (chlorophyll pigments). Here also are the enzymes for electron transfer and photophosphorylation. The second phase is the dark phase (2), during which there is synthesis of organic substances (carbon dioxide reduction) powered by the ATP and NADPH synthesized in the preceding phase. Phase 2 takes place principally in the stroma. Glucose formed is exported into the cytoplasm. After temporary storage there it is polymerized into starch.

5-34. Localization of photoreduction in a chloroplast. Spinach leaf, ×40,000.

The reaction detected here by cytochemistry is hydrogen transport. It takes place only in the presence of light. Reaction products form exclusively on internal membranes (th = thylakoids), which appear dense. The plastid envelope does not react. Granules present in the stroma are naturally dense.

5-35 and 5-36. Particular structure of chloroplast thylakoids. Spinach leaf, freeze-etched, ×95,000 and ×80,000.

Fractures that are transverse with respect to the thylakoids (th) have profiles quite comparable with those observed on ultrathin sections. Fractures at the surface or on the plane of membranes show constituent particles of different sizes. a = starch.

5-36

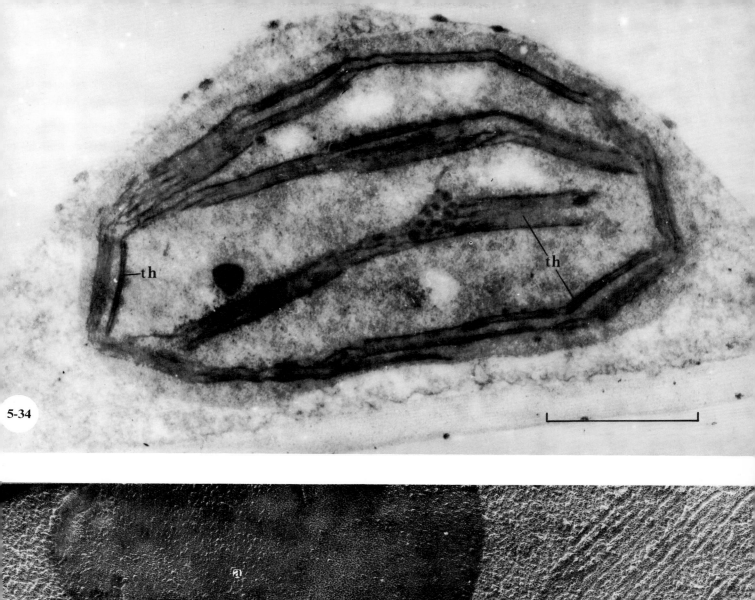

5-34

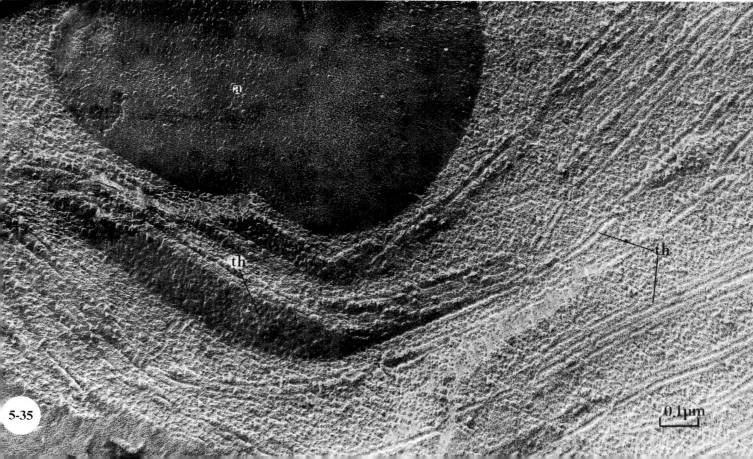

5-35

0.1µm

PLASTIDS: STORAGE

Under certain physiologic conditions, plastids do not form chlorophyll. Instead they are specialized for protein synthesis (proteoplasts) and for the synthesis of lipids and of pigments (chromoplasts). Most often they serve as storage organelles that concentrate starch (amyloplasts).

5-37. Amyloplasts in a storage tissue. Pea root parenchyma, ×25,000.

Starch is demonstrated through a reaction for polysaccharides. The internal membrane system is practically absent from these plastids, which are filled with bulky grains of starch (a) that are very reactive. m = mitochondria; w = wall.

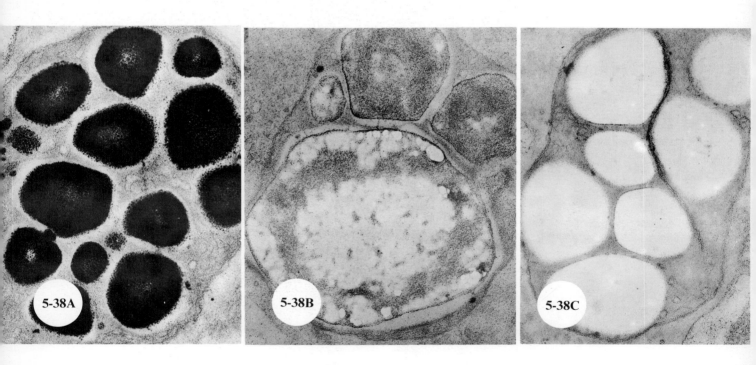

5-38. Characterization of starch by cytochemical test and by selective extraction. Same material as in 5-37, ×40,000.

In situ characterization of a product of cellular activity generally poses a difficult technical problem. In the case of starch it is relatively easy; it can be done by applying a cytochemical test which detects sites rich in polysaccharides (see p. 108). Digestion of a cell product by a specific hydrolase permits definition of its nature.

A. Polysaccharide test: positive reaction.

B. Treatment for three minutes with α-amylase, an enzyme which specifically hydrolyzes α, 1-4 glycosidic bonds. Partial extraction.

C. The same treatment prolonged for 10 minutes. Complete extraction.

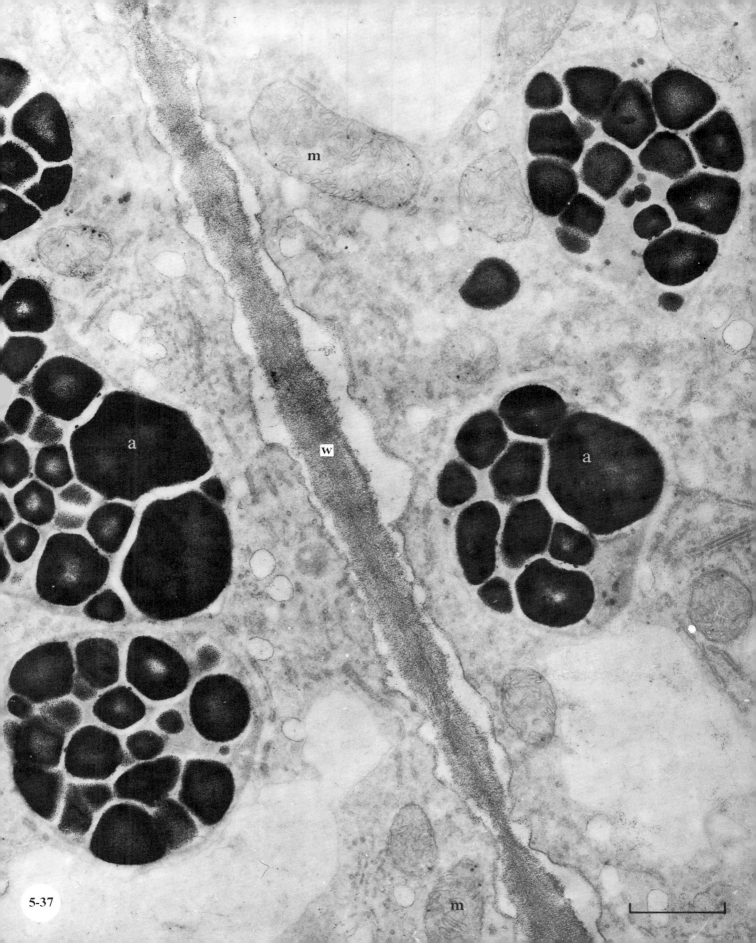

5-37

6. KINETIC AND CONTRACTILE APPARATUS

Centrioles, microtubules, and microfilaments are involved in the mechanical activities of supporting or moving the cell (contraction, cellular division, organelle migration, morphogenetic movements, cell movements in the external medium, or movement of the external medium around the cell).

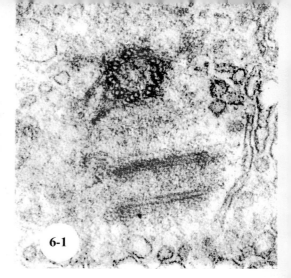

6-1

CENTRIOLES, CILIA, AND FLAGELLA

Centrioles generally occur in pairs. The two centrioles of a pair are perpendicular to each other, and each is made up of nine triplets of short tubules. Centrioles are present in nearly all animal cells and are located near the nucleus in a cytoplasmic area rich in microtubules and surrounded by Golgi bodies. They are at the poles of the mitotic spindle during cell division. In ciliated and in flagellated cells, they are often localized in the peripheral cytoplasm, where they constitute the base of cilia and of flagella (basal body). Cilia and flagella are formed by nine doublets of peripheral tubules and one axial doublet. Figures 6-1 and 6-2 show centrioles from a nerve cell.

6-1. A pair of centrioles. One is cut transversally, and the other longitudinally. ×60,000. (J. Taxi.)

6-2. Detail of a transverse section of a centriole. ×160,000.
Dense material covers the triplets of tubules, forming a peripheral crown (satellites).

6-3. Duplication of a centriole during cell division. Medusa cell, ×58,000.
Procentrioles form perpendicular to the two preexisting centrioles.

Figures 6-4 and 6-5 show cilia and basal bodies as seen in a ciliated protozoan.

6-4. Peripheral cytoplasm. ×35,000.
Cilia (c) and intracytoplasmic basal bodies (bb) are cut transversally at their distal region, where tubules are already organized in doublets. cm = cytoplasmic membrane.

6-5. Base of a cilium, longitudinal section. ×50,000.

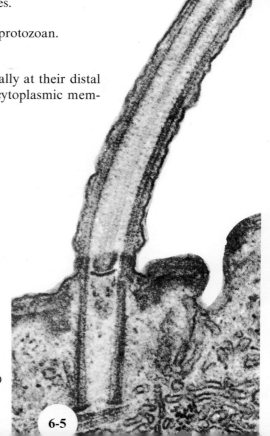

6-5

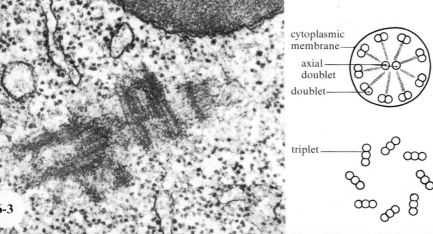

cytoplasmic membrane

axial doublet

doublet

triplet

6-3

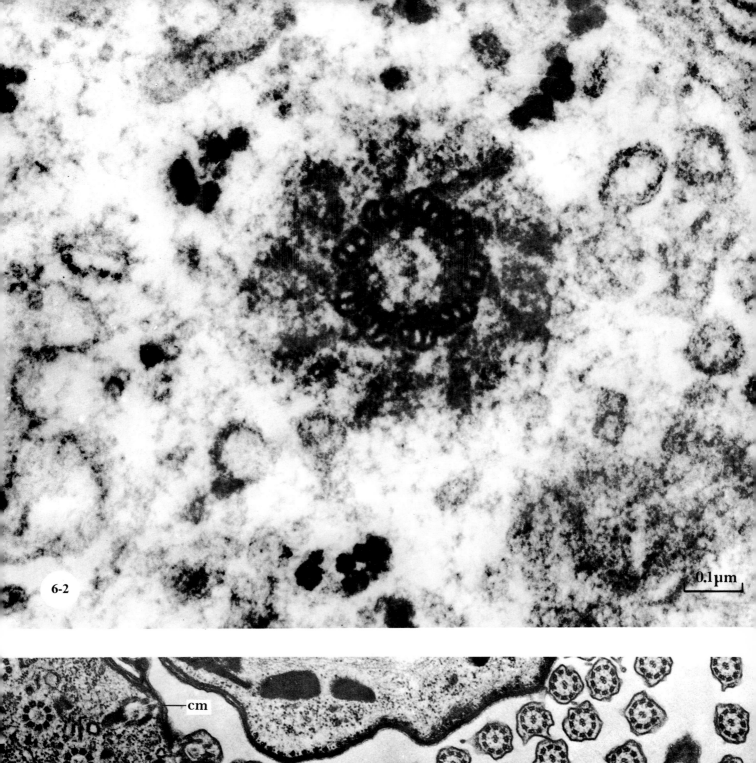

6-2

0.1μm

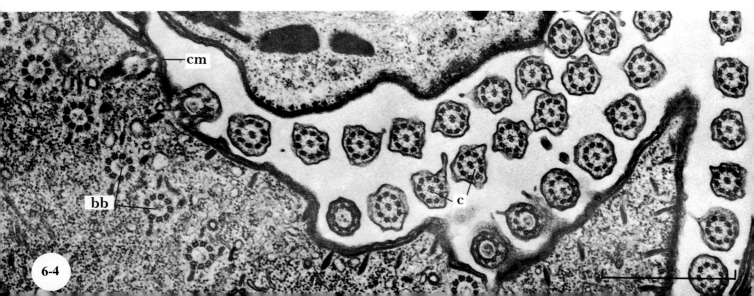

6-4

cm

bb

c

MICROTUBULES AND MICROFILAMENTS

Microtubules have diameters from 180 Å to 300 Å and are found in the cytoplasm, either dispersed, perhaps according to the orientation of cytoplasmic currents, or grouped in bundles. When grouped they sometimes form a veritable cytoskeleton. Microfilaments are finer units, having diameters of about 50 Å to 100 Å, and they play a role in the maintenance of cellular shape as filaments of the microvilli and desmosomes (see p. 14). Both of these structures are protein in nature. Figures 6-6, 6-7, and 6-8 show microtubular structure as seen in cilia of mussel gills. (F. Warner and P. Sater.)

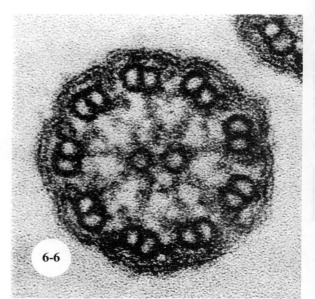
6-6

6-6. Cilium, transverse section. ×300,000.

6-7. Isolated microtubules, negative staining. ×300,000.
Each is formed by thirteen longitudinal protofilaments.

6-8. Demonstration at high magnification of the subunits constituting the wall of microtubules. ×620,000.

6-9. Interpretation of the structure of a tubule.
The microtubular wall is made up of globular protein units having a diameter of about 50 Å and arranged in a helix. A longitudinal array of globules make up the protofilament.

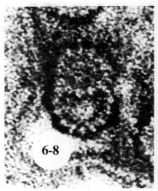
6-8

6-10. Microtubules as support structures. Cytopharynx of a protozoan, ×35,000. (C. K. Pyne.)

6-11. Microtubule of a flagellum. Locust spermatid, ×45,000.
Besides the microtubules of the flagellum, other tubules surround mitochondria. They have a morphogenic role, and they disappear when the cell stops elongating.

6-12. Microfilaments of the furrow of division. Segmenting Medusa ovum, ×45,000.
Microfilaments that carpet the internal face of the cytoplasmic membrane (cm) are involved in the progression of the cleavage. Two groups of microtubules of the spindle of division are cut transversely.

6-9

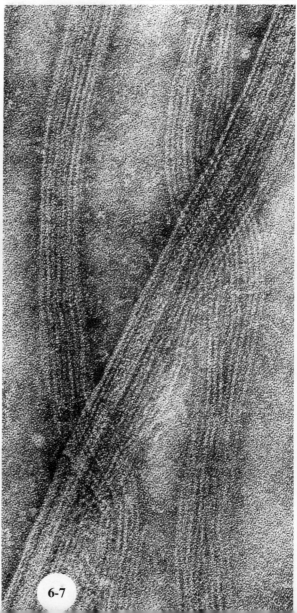
6-7

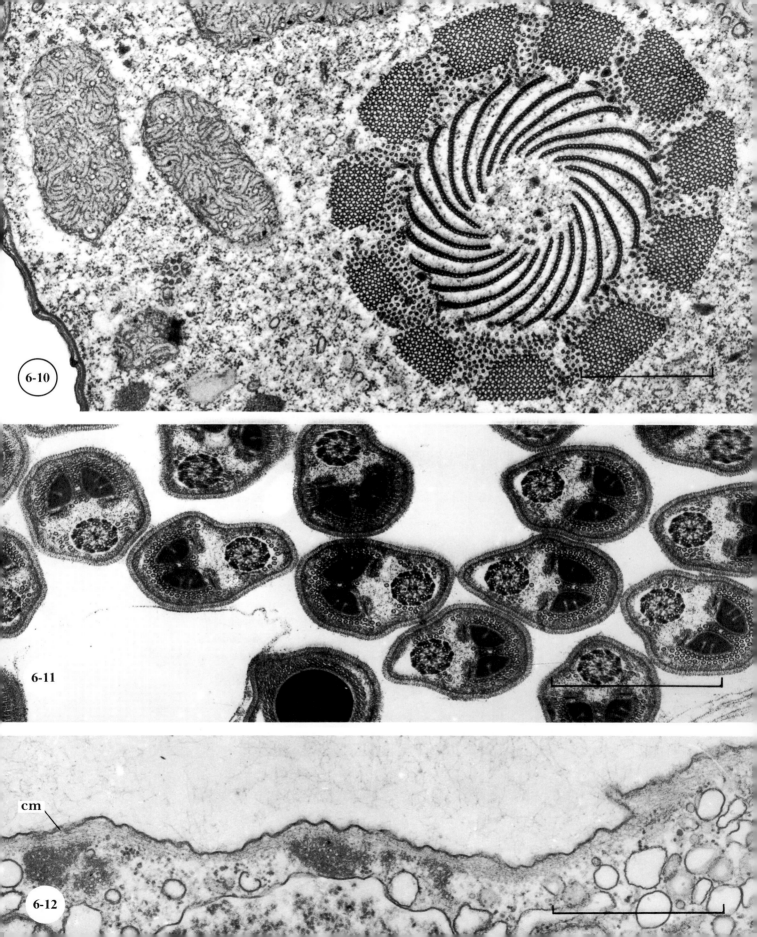

6-10

6-11

6-12

cm

MYOFILAMENTS

A striated muscle fiber is a giant, multinucleated cell that is rich in mitochondria and mostly occupied by myofibrils. The striated appearance of a myofibril results from the regular disposition of myofilaments which constitute it. Thick filaments (100 Å in diameter) contain myosin, and thin filaments (60 Å in diameter) contain actin. Actin and myosin are two contractile proteins.

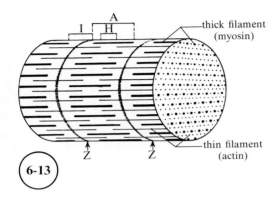

6-13. Organization of a myofibril.
Parallel, thick filaments form the dark bands, A, which are separated by clear bands, I. Thin filaments are attached to a dense transverse structure: the Z-line. The unit between the two Z-lines is a sarcomere (contraction unit).

Figures 6-14 and 6-15 show the disposition of myofilaments in a myofibril.

6-14. Transverse section of a myofibril. Wing muscle of a fly, ×100,000. (J. Auber.)
m = mitochondrion.

6-15. Longitudinal section of a myofibril. Gastrocnemius muscle of a batrachian, ×25,000. (J. Auber.)

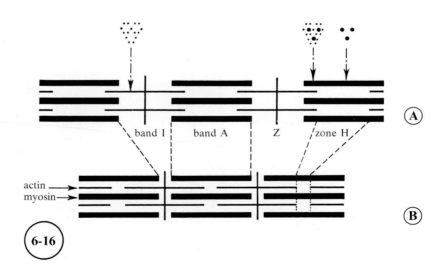

6-16. Shortening of sarcomeres during muscular contraction.
Contraction results from the telescoping of myofilaments, which glide over one another. The I-bands are thus shortened in contraction (B).

0.2 μm

m

6-14

6-15

7. CYTOPLASMIC INCLUSIONS

Products of cellular activity accumulate directly in the cytoplasmic matrix without being limited by a membrane. These products may be storage substances or they may be metabolic waste products. Some have a particular physiologic role, such as the respiratory proteins hemoglobin and hemocyanin. Figures 7-1 and 7-2 show glucidic stores in the cell.

7-1. Glycogen rosette in hepatic cell. Frog liver, ×50,000.
These glycogen units are 150 Å to 300 Å in size and are often associated with tubules of smooth endoplasmic reticulum. The tubules play a role in the storage of glycogen and in the mobilization of these stores.

7-2. Use of glucose in glycogen synthesis. Demonstrated by autoradiography of a human hepatoma, ×18,000. (S. Karasaki.)
The cell was incubated in a medium containing tritiated glucose. After autoradiography silver grains are seen, which indicate the sites of incorporation (black coils). They are localized exactly above an area of glycogen (gl). In normal conditions, blood glucose is accumulated in cells, where it is polymerized into glycogen. cm = cytoplasmic membrane; n = nucleus.

7-3. Lipid accumulation. Rat connective tissue cell, ×12,000.
Mitochondria are frequently associated with lipid globules. They play a role in their synthesis and in their mobilization.

Figures 7-4 and 7-5 show the accumulation of respiratory proteins with the example of hemocyanin in the blood cell (cyanocyte) of the limulus. Hemocyanin is a copper-containing respiratory pigment which is found in numerous mollusks and arthropods. It is made in the cytoplasm of blood cells in the form of bulky crystals.

7-4. Hemocyanin crystals. ×43,000.
The crystals are synthetized on polyribosomes (*arrow*) dispersed in the cytoplasm.

7-5. Hemocyanin molecules. Negative staining after dispersion of the crystal, ×210,000.
The basic unit of the crystal, the monomer, is a globular protein that is 75 Å in diameter and has a molecular weight of 500,000 daltons. The monomers are often associated in dimers and tetramers, seen here face-on and in profile.

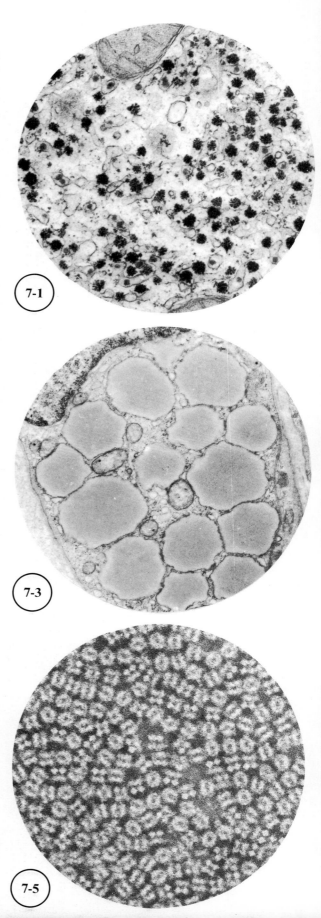

7-1

7-3

7-5

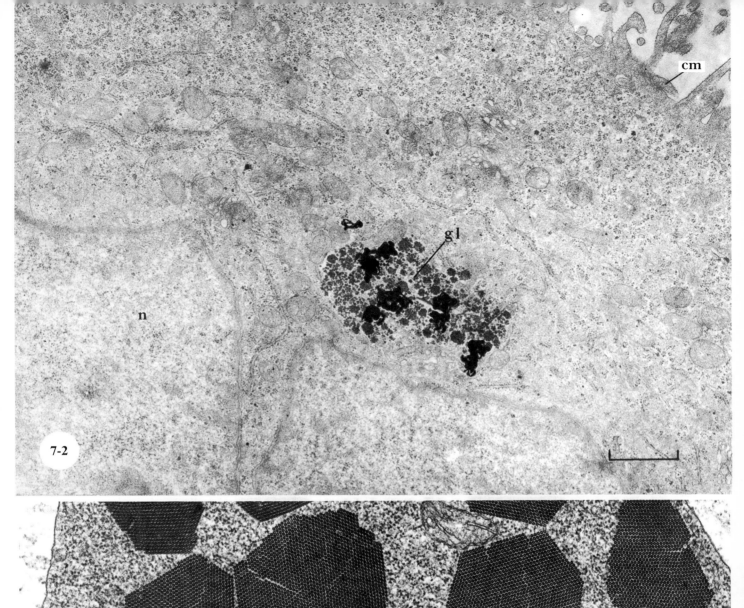

8. NUCLEUS

The nucleus contains the genetic information of the cell in the form of DNA. It is the site of RNA transcription: messenger RNA, transfer RNA, ribosomal RNA are made here and transported to the cytoplasm. The nucleus contains chromatin (despiralized chromosomes), which is made up of DNA bound to basic proteins (histones), and nucleoli, which are rich in RNA. Chromatin and nucleolus are bathed in nucleoplasm. Topographic segregation of nuclear functions is characteristic of the eucaryotes, in which the nucleus is separated from the cytoplasm by a nuclear envelope formed by two membranes.

AUTOCATALYTIC AND HETEROCATALYTIC ACTIVITY OF THE NUCLEUS IN INTERPHASE

Heterocatalytic activity of the nucleus is the function of the genetic material: RNA transcription. Autocatalytic activity is the reproduction of the genetic material: duplication of DNA. These two activities take place during interphase, the interval separating two cellular divisions.

8-1. **Morphologic features of the interphase nucleus.** Blood vessel of rat embryo, ×10,000.
Chromatin is dispersed in the nucleus of epithelial cells, A. Chromosomes are in an unwound state; the cell is active. In cell B (a future red blood cell), chromatin is condensed in places (*arrows*), which indicates a lower order of activity. en = nuclear envelope; nu = nucleolus.

8-2. **DNA synthesis in the nucleus.** Follicular cell of a butterfly ovary after autoradiography, ×15,000. (M. Guelin.)
A specific radioactive precursor of DNA, ^{3}H-thymidine, was injected into the animal. In autoradiographs of thin, sectioned cells, silver grains are seen to be localized selectively over the chromatin. They indicate the sites where DNA is being synthesized. This period of autocatalytic activity occurs during a specific part of interphase (the synthesis or S-phase). nu = nucleolus; en = nuclear envelope.

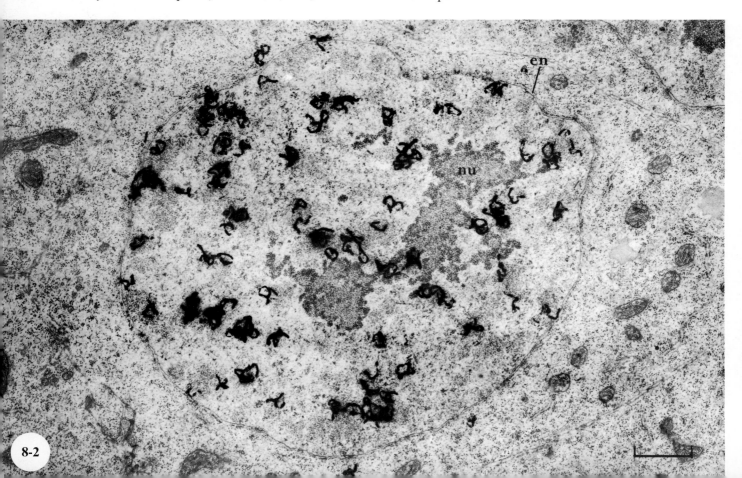

8-2

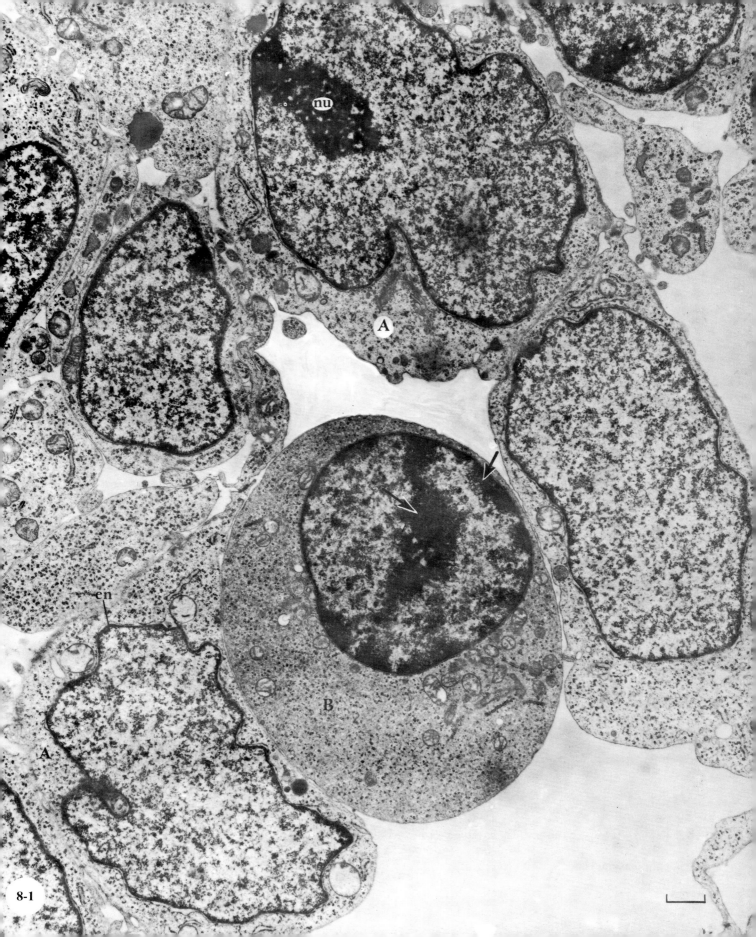

nu

A

A

en

A

B

8-1

NUCLEAR ENVELOPE

The nuclear envelope which characterizes the nucleus of eucaryotes is a differentiated portion of the endoplasmic reticulum. It is composed of two membranes delimiting a perinuclear space which, in places, is in continuity with cavities of the reticulum. It is interrupted by pores. It is a border which controls nucleocytoplasmic exchanges, including the flow of small molecules and of ions and the passage into the cytoplasm of macromolecules, particularly ribonucleoproteins of the nucleus.

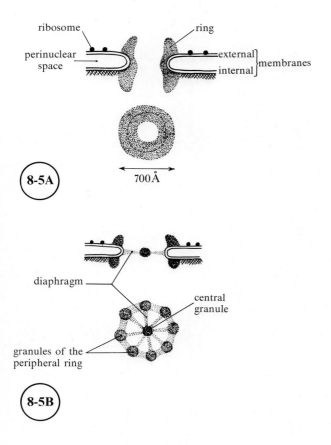

8-3. Nuclear membrane. Surface view after freeze-etching. Pea root, ×50,000.

Nuclear pores (np) are numerous here; they represent about 25% of the nuclear surface. The number and disposition of the pores vary with cellular activity.

8-4. The two membranes of the nuclear envelope. Human oogonium, ×65,000.

In the nucleoplasm, some particles seem to be entering the pore (np), in which there is a dense structure, possibly a diaphragm. inm = internal nuclear membrane; enm = external nuclear membrane; pns = perinuclear space; ch = chromatin; rb = ribosomes.

8-5. Interpretation of pore structure.
The presence of a diaphragm in the pore is controversial.
A. Open pore with a ring of dense material.
B. Pore with a diaphragm.

8-6. Reconstitution of the nuclear envelope. Human oogonium, ×22,000.
After cellular division, the nuclear envelope (en) re-forms around the chromosomes (chr) from fragments of endoplasmic reticulum. Nuclear pores will form later. *Above* is an interphase nucleus (n); cm = cytoplasmic membrane.

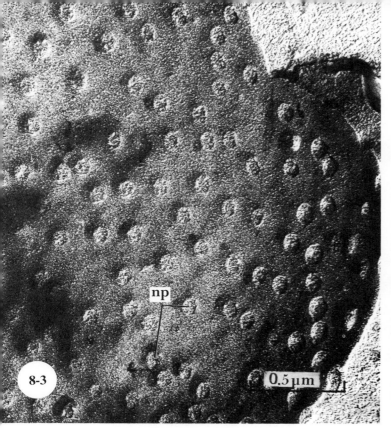

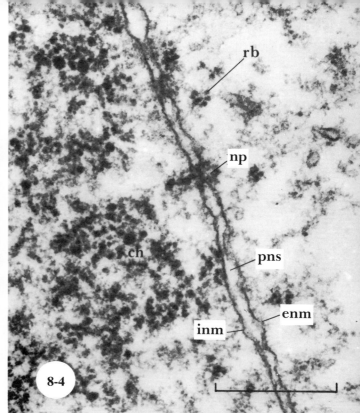

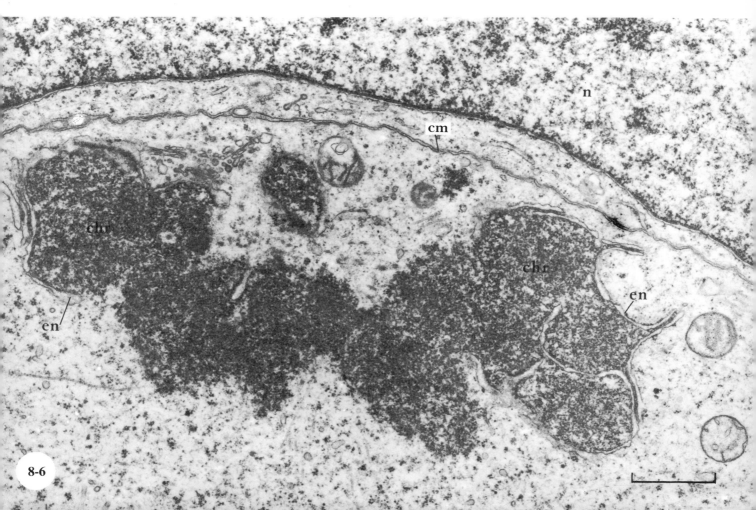

NUCLEOLI

Nucleoli are the site of ribosomal RNA synthesis. They are often associated with portions of chromatin, where they originate. Their size and, in certain cases, their increased numbers (in the oocyte, for example) betray an intense cellular activity.

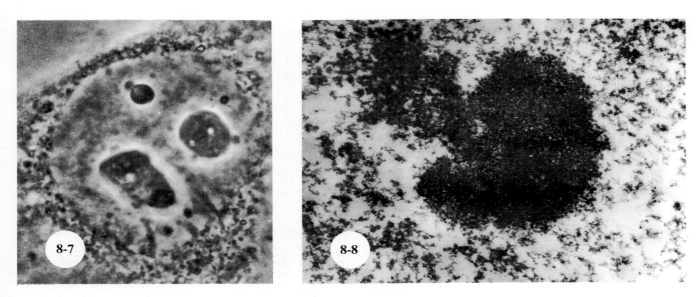

8-7. **Nucleolus of a living cell.** Mouse fibroblast in culture seen by phase-contrast microscopy, ×2000.
Nucleoli appear as three unequal masses that are dense and refringent.

8-8. **Chromatin associated with the nucleolus.** Pea root, ×35,000.
The chromatin associated with the nucleolus (*upper left*) probably corresponds to the chromosomal region observed in contact with the nucleolus at the beginning of cell division (the nucleolar organizer).

Figures 8-9 and 8-10 show the fibrillar and granular structure of the nucleolus as seen in rat embryo cells. The structural duality of a nucleolus is particularly visible in embryonic cells.

8-9. **Eight-cell stage.** ×20,000.
The central part is made up of fibrillar units from 60 Å to 80 Å in diameter that form a compact mass (*). At its periphery, this mass is reduced to cords, with which are associated granules 150 Å to 200 Å in diameter (**). en = nuclear envelope.

8-10. **Sixteen-cell stage.** ×27,000.
The fibrillar portion is reduced, and this reduction becomes accentuated during ensuing divisions. The fibrillar portion fragments and becomes more difficult to see. Note that the nucleolus is not limited by a membrane.

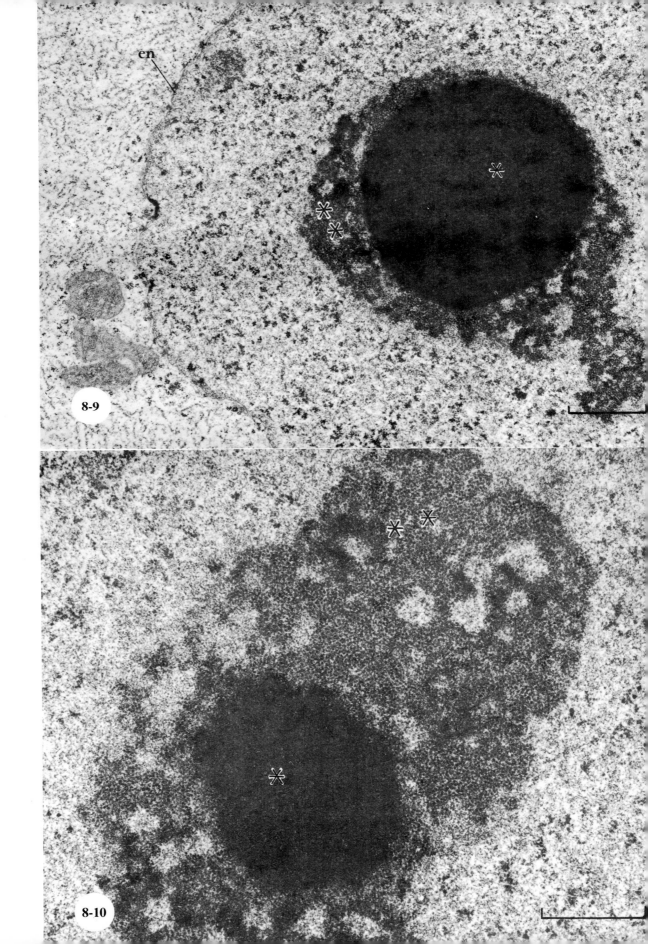

en

8-9

8-10

TRANSCRIPTION OF RIBOSOMAL RNA

Information contained in nuclear DNA in the form of the nucleotide sequence (genes) is reproduced base to base in a molecule of RNA during transcription.

8-11. Transcription of a sequence of nucleotides.
A, C, G, and T are the four bases of DNA.
A, C, G, and U are the four bases of RNA.
A = adenine;   C = cytosine;   G = guanine;   T = thymine;
U = uracil.
One of the threads of the DNA double helix (here, the one on the right) codes an RNA molecule.

```
A = T        A
C ≡ G        C
T = A        U
G ≡ C  →     G
G ≡ C        G
A = T        A
T = A        U

 DNA         RNA
```

8-11

8-12. Cytologic aspect of the transcription of ribosomal genes. Triton oocyte, ×23,000. (O. Miller and B. Beatty.) The first electron microscope demonstration of the phenomenon of transcription was similar to this study. Synthetic activity in the oocyte is intense. Its nucleus is giving rise to large quantities of messenger RNA and ribosomal RNA (rRNA) which will accumulate in the cytoplasm. There are many nucleoli in which ribosomal genes are repeated several times.

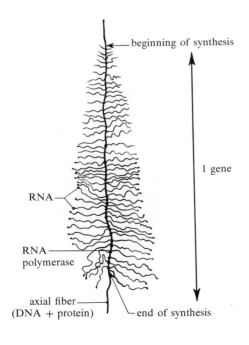

Under certain conditions of isolation, the fibrillar material of nucleoli spreads out. It consists of an axial fiber (af) covered at regular intervals by unit matrices (um), which are composed of ribonucleoproteins. DNA segments covered by a unit matrix represent the ribosomal genes, and the matrix filaments are ribosomal RNA molecules transcribed on the gene. Numerous RNA molecules are simultaneously transcribed on each gene, and their increasing length indicates the stage of transcription. This gradual increase in the length of the transcribed RNA gives the unit matrices a plume-like shape. RNA polymerase is situated at the base of each forming RNA molecule.

Considerable quantities of RNA are synthesized, since the genes are repeated along the DNA fiber. RNA associated with proteins forms the granular part of the nucleolus. The granules later move to the cytoplasm, where they give rise to ribosomes.

8-13

8-13. RNA transcription.

um

af

8-12

8-14. **Gene function in giant chromosomes.** Salivary glands of an insect larva (chironome), ×900. (W. Beermann.) In the nucleus of salivary glands, several DNA replicative cycles succeed each other without separation of the daughter chromosomes. The giant chromosomes, or polytenes, are bundles of interphase chromosomes; moreover, the homologous chromosomes are associated (*arrow*). The dark bands are aligned spiralized portions of chromosomes and are rich in DNA.

When despiralization occurs in these dark bands, "puffs," or rings of Balbiani (br), appear upon which one may show RNA synthesis. This is messenger RNA. The dark bands, thus, are genes, and the "puffs" are these genes in function. The distribution of the puffs varies during the lifetime of the cell, which indicates that there is a modulated functioning of genes along the chromosomes. Some hormones can specifically activate certain puffs and thus influence cell function at the highest level, that of transcription. nu = nucleolus.

8-15. **Gene function in the lampbrush chromosomes.** Triton oocyte seen by phase-contrast microscopy, ×900. (J. Gall.) A pair of homologous chromosomes was isolated from a nucleus in the prophase of meiosis. At this stage, chromosomes have lateral expansions in the form of loops, giving rise to their lampbrush appearance. Enzymatic digestion and autoradiography led to the interpretation that these loops are sites of messenger RNA synthesis, and electron microscopic examination revealed features similar to those shown in Figure 8-12. Thus the lampbrush configuration reflects an intense activity of the genes.

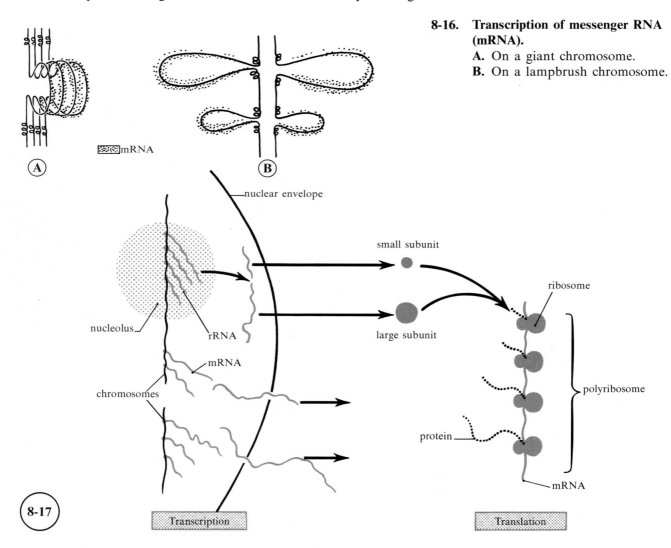

8-16. **Transcription of messenger RNA (mRNA).**
A. On a giant chromosome.
B. On a lampbrush chromosome.

8-17. **Transcription and translation.**

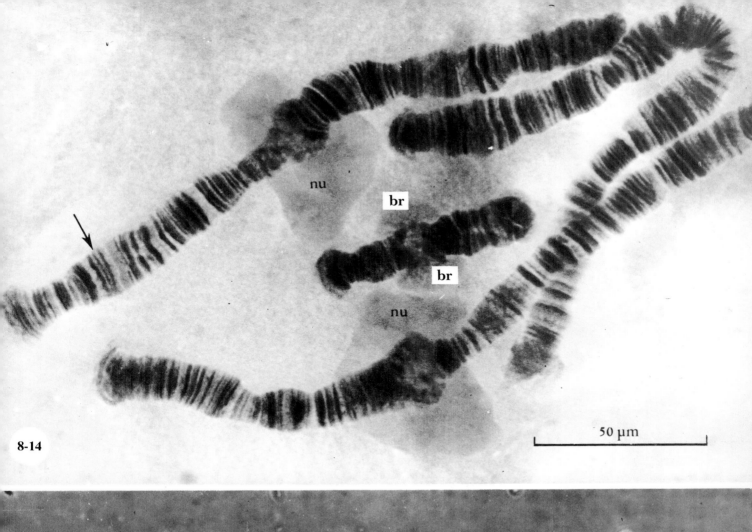

nu
br
br
nu

50 µm

8-14

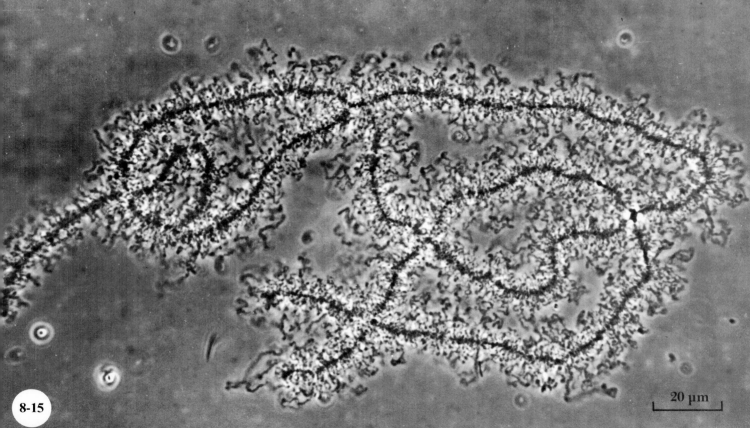

20 µm

8-15

CHROMOSOMES AND KARYOTYPES

Diploid cells (2n) have a stock chromosome of maternal origin (n) and a stock chromosome of paternal origin (n). In many species one may distinguish autosomes, which are present as morphologically identical pairs, and a pair of heterosomes (sex chromosomes), which are different in the two sexes. Thus the human species is characterized by 46 chromosomes, of which 22 pairs are autosomes and 1 pair is sex chromosomes. Figures 8-18 to 8-20 illustrate the set of human chromosomes. (J. De Grouchy.)

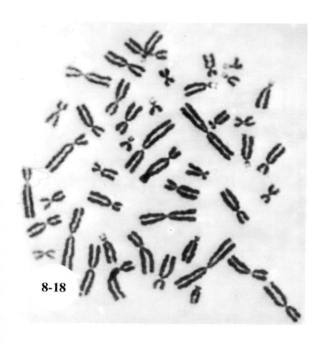

8-18.

8-18. Chromosome display. Somatic cell in culture, ×800. Cell division was blocked in metaphase by colchicine. At this stage the daughter chromosomes are still joined by the centromeres.

8-19. The 22 pairs of the karyotype. ×2500.
The karyotype is obtained by classifying the chromosomes in seven groups according to an international convention. Their size and the position of their centromere are used for this classification.

8-20. Sex chromosomes X and Y. ×2500.
A. The female has two morphologically identical chromosomes, XX.
B. The male has two different chromosomes, X and Y.

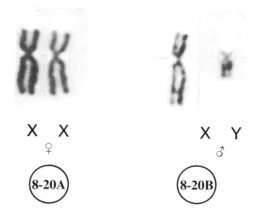

X X
♀

X Y
♂

8-20A   8-20B

8-21. Chromosome 12 of the human species. ×50,000. (E. Dupraw.)
A metaphase chromosome as seen by electron microscopy. The picture shows clearly the two chromatids, the position of the centromere at the point of constriction, and the fibrillar structure of the chromosomal material. It is thought that the chromatid is an extremely long and convoluted fiber 200 Å to 300 Å in diameter. The mode of association between DNA and proteins in this fiber remains hypothetical, and the existence of subunits is controversial.

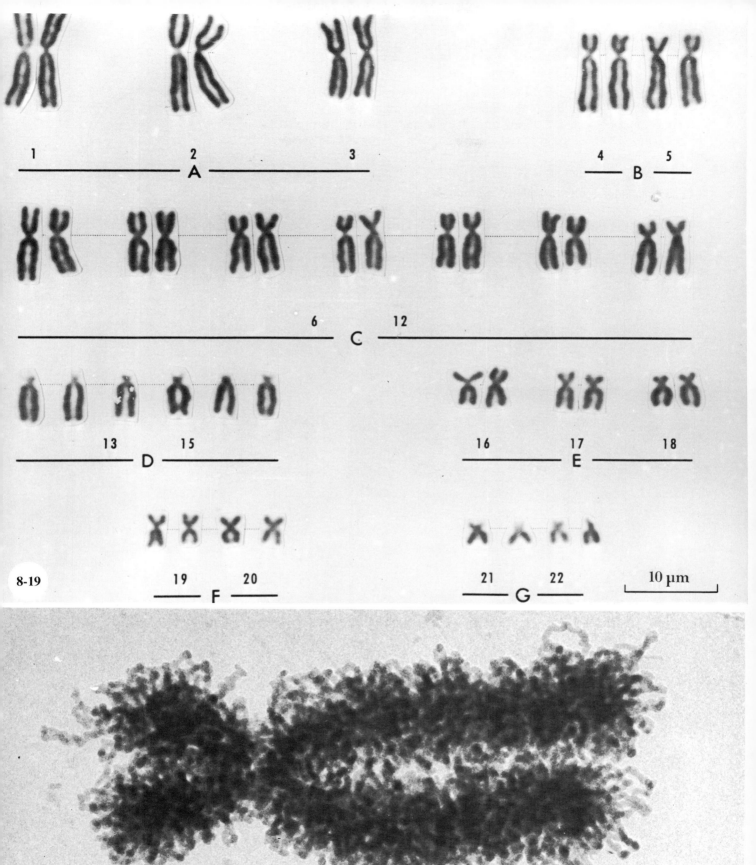

1 2 A 3 4 B 5

6 C 12

13 D 15 16 E 17 18

19 F 20 21 G 22 10 µm

8-19

8-21

9. CELLULAR DIVISION

Eucaryote cells divide by mitosis. During the course of mitosis, genetic material of the chromosomes is distributed equally to the two daughter cells. Replication of this material (autocatalytic activity) has taken place beforehand, during the cell cycle (see p. 66). During mitosis, chromosomes become condensed and their heterocatalytic activity is reduced.

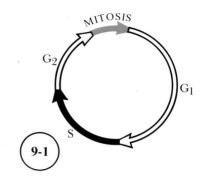

9-1. The cellular cycle: interphase and mitosis.
G_1 = presynthesis phase; S = phase of DNA synthesis; G_2 = postsynthesis phase.

9-2. Variation in DNA content of the nucleus during the cell cycle.

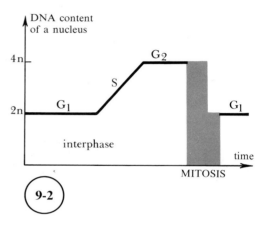

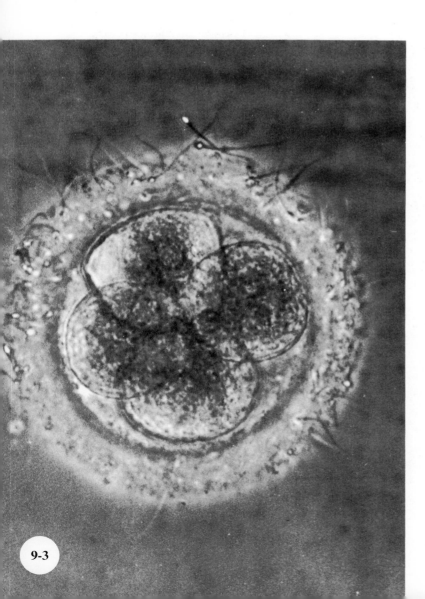

9-3. Morphologic aspects of cell division. ×800. (R. Edwards.)
A human egg after fertilization in vitro. The first two divisions have taken place. The four embryonal cells (blastomeres) are protected by a layer of mucopolysaccharides (zona pellucida), in which numerous spermatozoons are retained.

9-4. Chromosome condensation during prophase of mitosis. Follicular cell of rat ovary, ×14,000.
Migration of a pair of centrioles (cs) has occurred. A pair of centrioles is placed on either side of the nucleus (only one centriole of each pair is visible in the picture). The nuclear envelope (en) is profoundly indented in the plane of the centrioles. The nucleolus has disappeared. Chromosomes undergo condensation while each is in contact with the nuclear envelope.

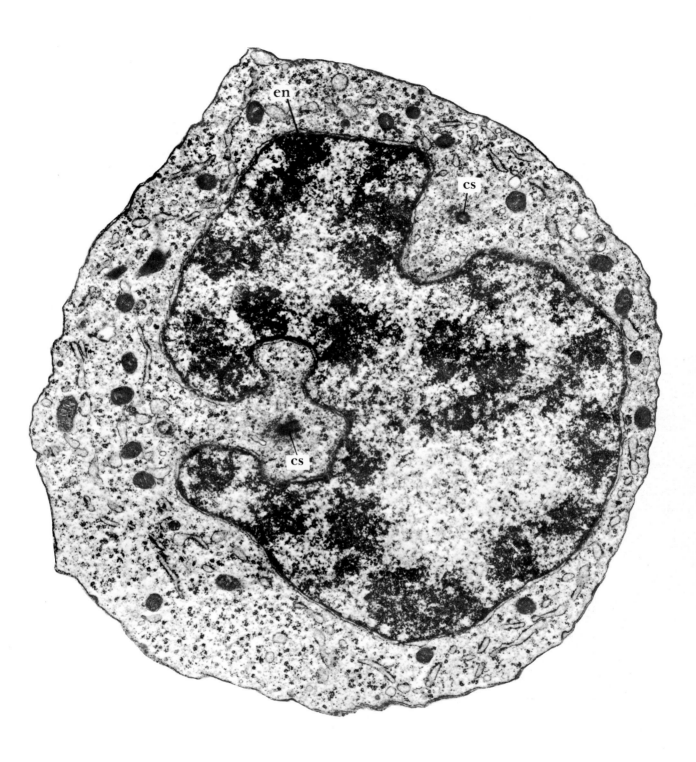

9-4

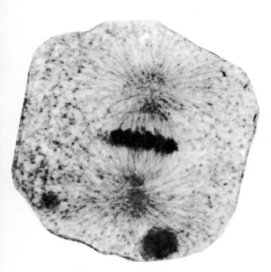

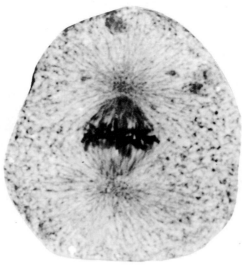

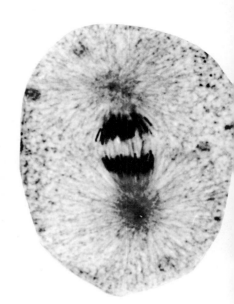

metaphase
(equatorial plane)

beginning of anaphase
(separation of the two
stock chromosomes after
cleavage of the centromeres)

end of anaphase
(migration toward the poles)

 9-5A

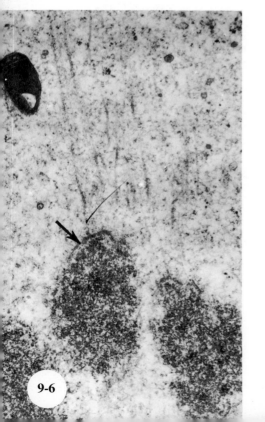

9-6

9-5. **Metaphase and anaphase of mitosis.** Segmenting fish embryo, ×200.
The cells are in different stages of division and are oriented along
different planes. The spindles (bundles of microtubules) and astrospheres
(radiating arrangement of microtubules around the centrioles) are readily
seen. Note particularly the equatorial planes of chromosomes in metaphase, seen either in profile or in a polar view.

9-6. **Attachment of spindle microtubules to the chromosomes during metaphase.** Mouse embryo, ×13,000.
Spindle microtubules are attached to the centromere, or kinetochore,
which is the dense material (*arrow*).

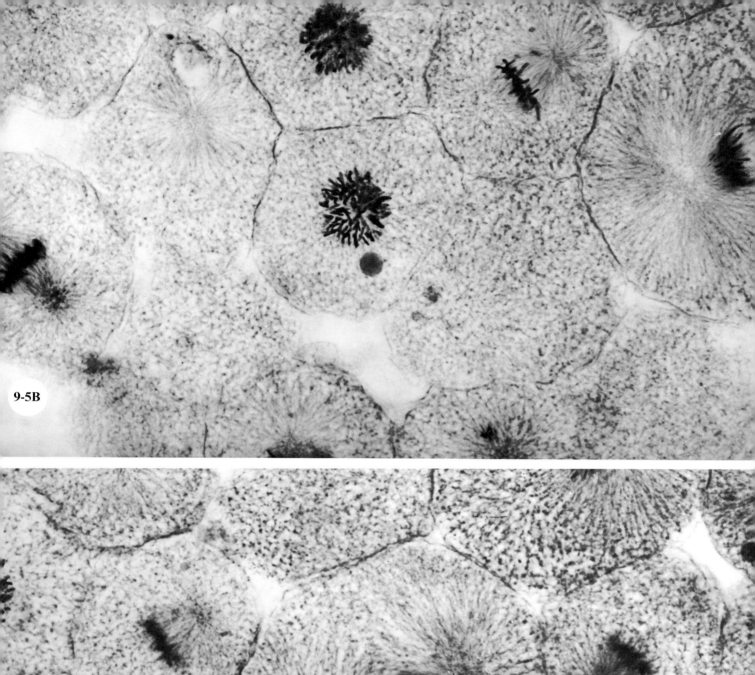

9-5B

9-5C

100 μm

9-7. Movement of the daughter chromosomes during anaphase of mitosis. Mouse embryo, ×14,000. During the polar migration of the chromosomes (chr), their centromeres are connected with the poles by microtubules (mt). Microtubules converge toward the centriole (cs) without having direct contact with it.

9-8. Reconstitution of the daughter nuclei during telophase of mitosis. Mouse embryo, ×12,000. The chromosomes have fused into two masses of dense chromatin (ch) in the vicinity of the poles. The nuclear envelope (en) re-forms. The cell grows in length, and its cleavage in the equatorial region (*arrows*) begins. m = mitochondrion.

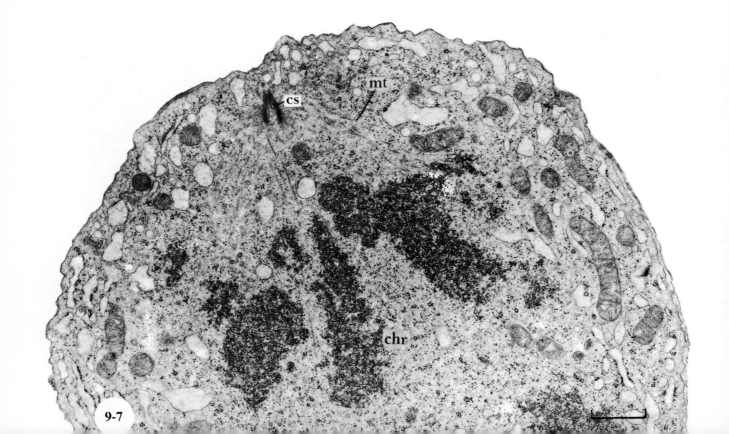

9-7

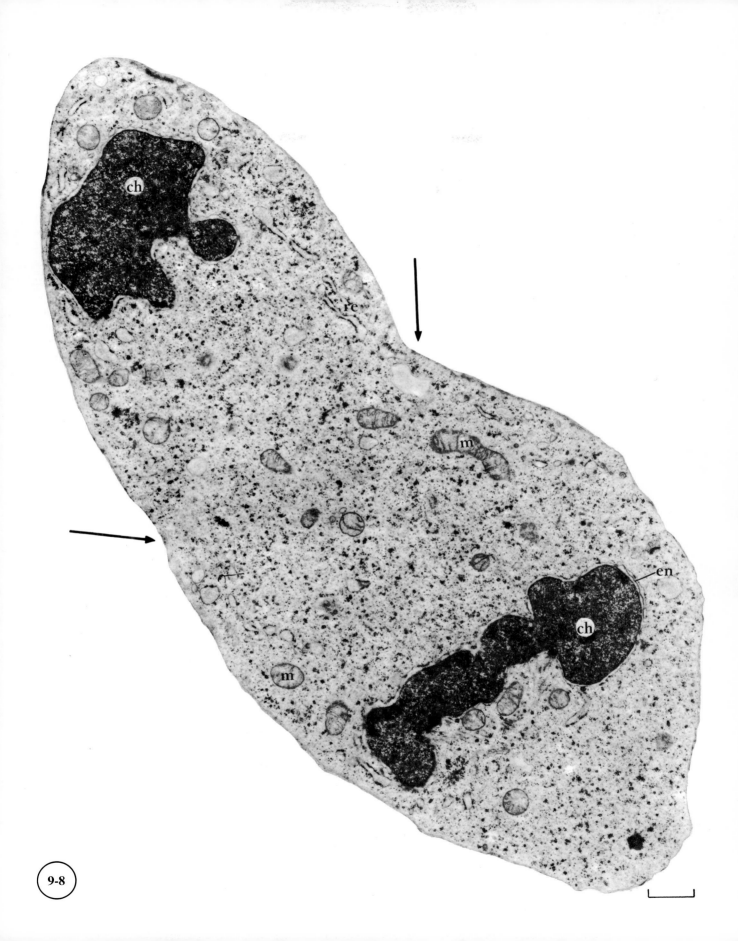

9-8

SEPARATION OF THE DAUGHTER CELLS

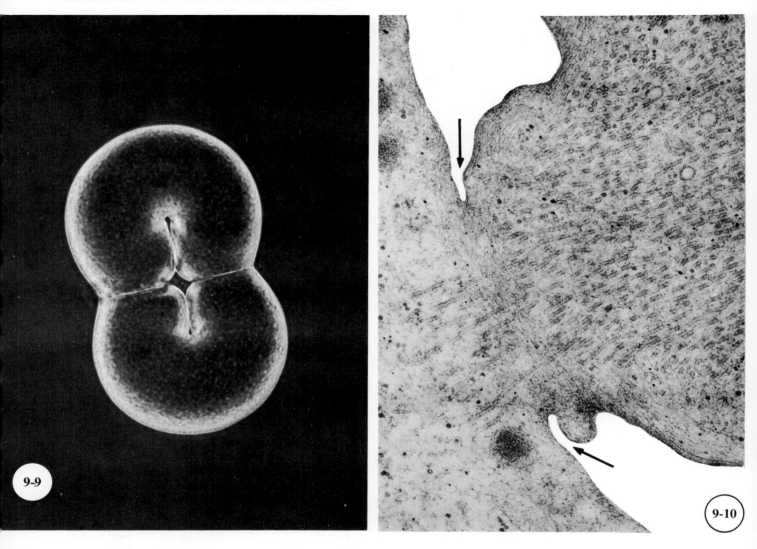

Figures 9-9 and 9-10 show cytokinesis of an animal cell.

9-9. **Segmentation.** Medusa ovum, ×250.
The first division has taken place. The fissure of the second division is deepening (in this species, fissure extension is asymmetric).

9-10. **Segmentation.** Rat embryo, ×35,000.
The section tangential to the dividing groove shows the microtubules of the spindle. In a perpendicular plane it shows very fine, contractile filaments (50 Å to 70 Å in diameter) which play a role in the extension of the fissure.

Figures 9-11 and 9-12 show partitioning of a plant cell.

9-11. **Phragmoplast.** Wisteria root, ×18,500. (E. Newcomb and W. Wergin.)
In the space contained between the two daughter nuclei, n, of the phragmoplast, a new wall forms by the coalescence of vesicles which reassemble in the equatorial plane. cm = cytoplasmic membrane; mt = microtubules; en = nuclear envelope; g = Golgi apparatus.

9-12. **Formation of the dividing membrane.** Bean root, ×50,000. (E. Newcomb and W. Wergin.)
The developing wall progresses away from the center to fuse with the wall of the mother cell. The vesicle membranes fuse and give rise to the new cytoplasmic membrane (cm). Communications persist between the daughter cells. They are the future plasmodesmata. mt = microtubules.

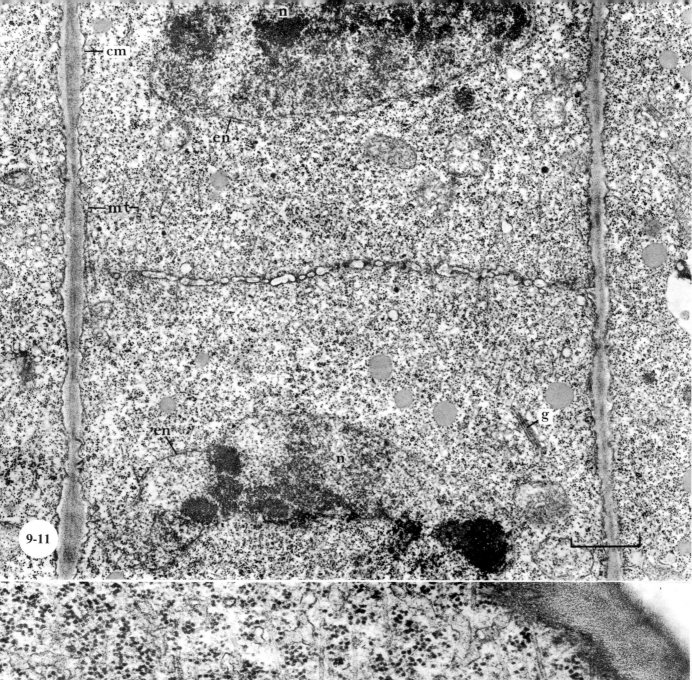

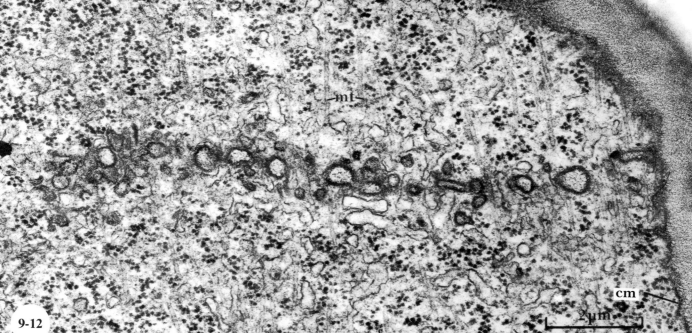

10. EXTRACELLULAR STRUCTURES

Extracellular structures may constitute a diffuse covering or a compact one around the cells. They comprise a basic hydrated substance, or matrix, and formed macromolecular elements such as collagen in animal cells and cellulose in plant cells. Enzymes (hydrolases) and cations (Ca^{++}) can be located there. These structures are involved in the relationships of the cell with the surroundings; they provide support, protection, selective filtering, and antigenicity.

Figures 10-1 and 10-2 show collagen of connective tissues in two animal structures. Collagen is a protein molecule forming long fibers and having a periodic structure; it is secreted by a connective tissue cell. The matrix which clothes it is rich in acid mucopolysaccharides. It is particularly well developed in cartilaginous tissues.

10-1.   **Connective tissue collagen.** Tracheal cartilage of the mouse, ×6500. (R. Seegmiller, C. Ferguson, and H. Sheldon.)

10-2.   **Connective tissue collagen.** Sigmoid valve of pig, ×25,000. (L. Zylberberg.)
n = nucleus; cm = cytoplasmic membrane. The insert on the margin shows negatively stained collagen fibers. Note their periodicity. (×150,000.)

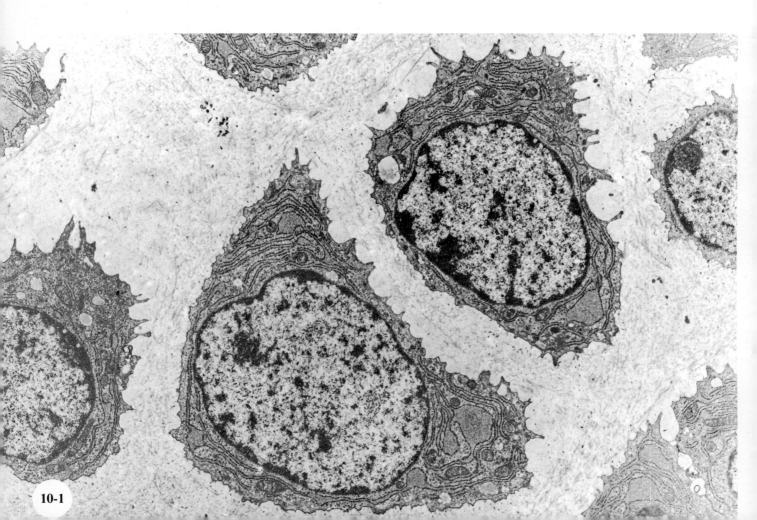

10-1

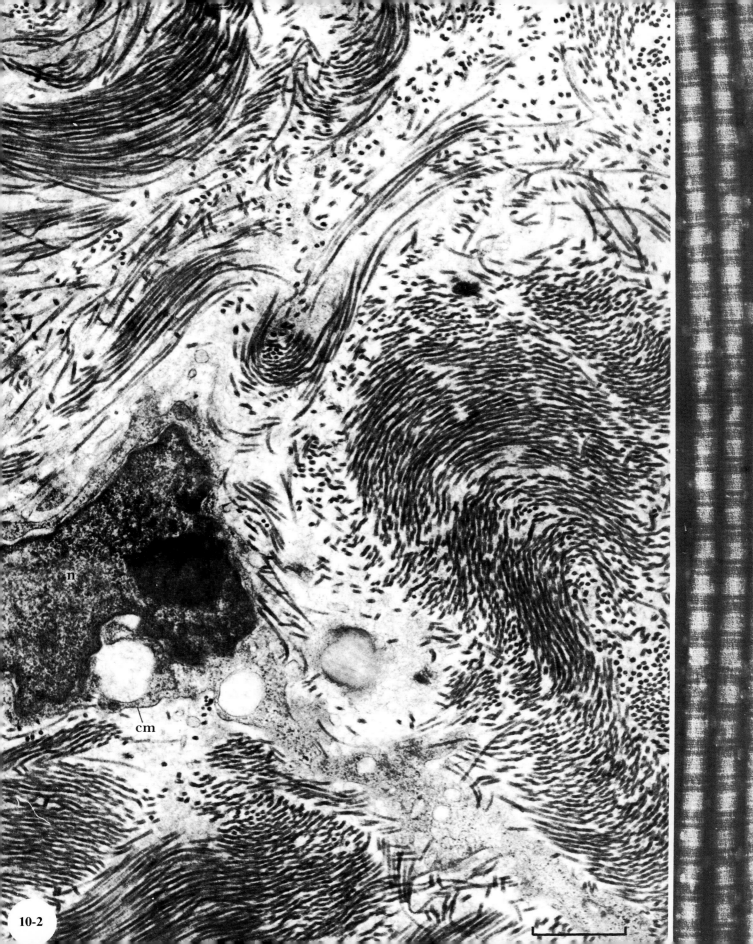

10-2

CELLULOSE AND WALL

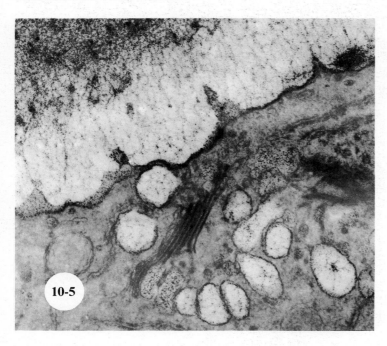

10-5

The wall is a nearly constant structure of plant cells. Cellulose forms its resistant framework. Cellulose is made from long chains of glucose arranged in a crystalline manner in unit fibrils. These fibrils are enshrouded within an amorphous matrix of neutral or acid polysaccharides, and of glycoproteins.

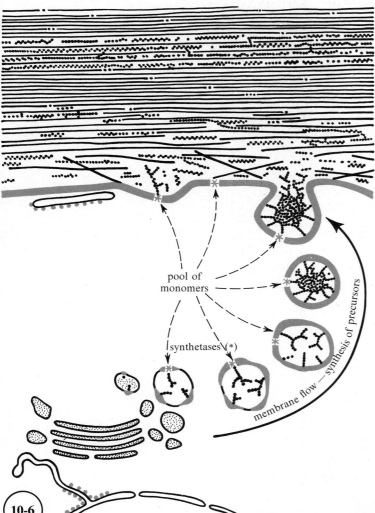

10-6

10-3. Wall of a plant cell. Pea root, ×45,000.
A test for polysaccharides shows the finely fibrillar texture of the cellulose-rich regions. The densest lamella, the middle layer, is formed by acid polysaccharides (pectins), which constitute an intercellular cement. lm = middle lamella; cm = cytoplasmic membrane.

10-4. Cellulose framework of the wall. ×30,000.
Shadowing after dissolution of the matrix reveals fibers in bundles of unitary fibrils. Their arrangement in a meshwork makes evident the position of plasmodesmata (pm), which are cytoplasmic bridges between neighboring cells.

10-5. Synthesis of wall constituents. Test for polysaccharides, ×30,000.
Wall constituents are elaborated in Golgi vesicles and on the cytoplasmic membrane.

10-6. Biosynthesis and secretion of wall constituents.
Certain membranes are adapted (membrane capacitation) and have the enzymatic equipment in the form of synthetases (* in the diagram) for polymerizing monomers (-hexoses, phosphate-hexoses, and nucleotide-hexoses). After membrane flow, the constituents (matrix and fibrils) are secreted by exocytosis. They associate with each other and form a vast web which surrounds the cell.

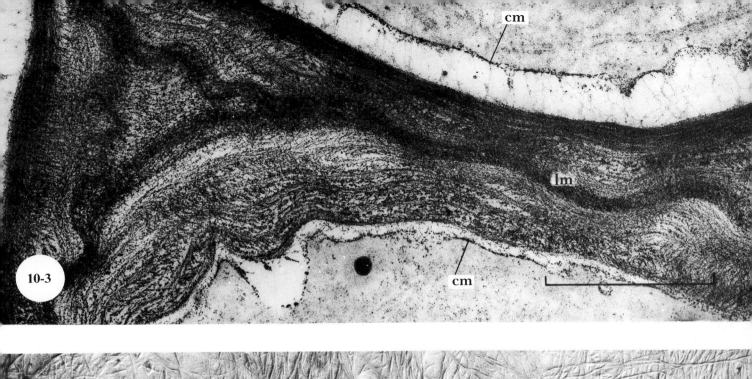

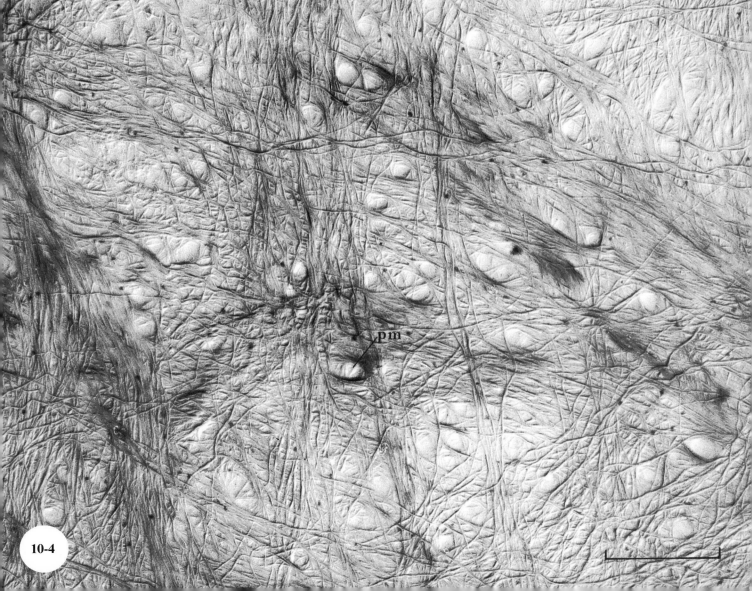

INTEGUMENTARY STRUCTURES

Cells directly in contact with the external medium are often protected by a thick, resistant covering. Such is the case in the cuticle of insects and in the envelopes of cells that are to be disseminated, such as pollen and spores. Use of the scanning electron microscope is the best method for studying the surface of these layers. Figures 10-7 and 10-8 are studies of insect cuticle.

10-7. **Transverse section of insect cuticle.** Abdomen of a larva of *Tenebrio* (mealworm), ×5000. (J. Delachambre.)
Cells of the epidermis secrete a stratified cuticle that is renewed at each molting.

10-8. **Surface of insect cuticle.** Eye of *Drosophila* under the scanning electron microscope, ×2000.
Facets of the compound eye have a geometric disposition. At each facet, the cuticle forms a transparent lens; while between the facets, short bristles emerge.

10-9. **Envelope of a pollen grain.** *Stellaria* under the scanning electron microscope, ×3000. (F. Roland Heydacker.)
The raised features are either ornaments or correspond to opercula (*arrow*) which allow the pollen tube and male nuclei to emerge. The shape and distribution of the ornaments are specific and are used in classification.

10-7

10-8   10 μm

10-9   10 μm

11. PROCARYOTES

Procaryotes have a relatively simple structure. Their genetic material is carried by a unique chromosome that is made up of DNA not bound to proteins. This chromosome is not isolated within a nucleus. Cellular compartmentation and membrane specialization are rudimentary. Notwithstanding their simplicity, however, procaryotes are typical autonomous cells able to elaborate the thousands of enzymes needed for their metabolism.

Apart from their biologic interest (they have extraordinarily varied life-styles), they have become a remarkable research tool for molecular biology. Their relatively primitive yet highly diversified structure confers on them an evolutionary interest of the first order. Procaryotes include bacteria, mycoplasmas, and the blue-green algae.

BACTERIA

Bacterial cells (of about 1 μm in size) have a resistant wall containing complex glycoproteins that are involved in immune and pathogenic phenomena. The cytoplasmic membrane bears numerous enzymatic and multienzymatic complexes which act in syntheses and in respiration. These complexes bear witness to the high degree of organization these cells have at the biochemical level. A membrane invagination forms the mesosome, which is associated with the chromosome and undoubtedly plays a role in its replication. The cell contains thousands of ribosomes and pooled metabolites. Division is very rapid (one or more per hour), and it occurs by transverse division.

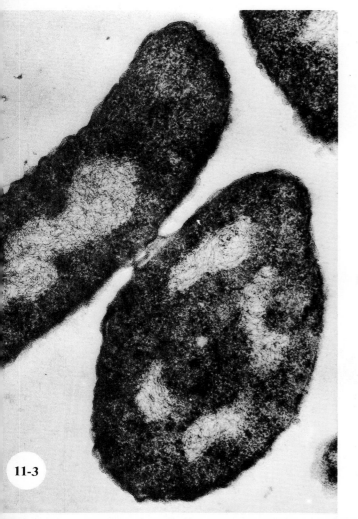

11-3

11-1. Structure and division of a bacterial cell. *Mycobacterium phlei,* ×80,000. (A. Petitprez and J. C. Derieux.)
Genetic material is represented by a coil of DNA fibrils (*arrows*). After division, these bacterial cells remain joined in chains. w = bacterial wall; cm = cytoplasmic membrane; ms = mesosome; ts = transverse septum; pr = polysaccharide reserves; rb = ribosomes.

11-2

11-2. Enzymatic complex isolated from *Escherichia coli.*
×1,000,000. (C. Gordon.)
This is a photomicrograph of glutamine synthetase, an enzyme having a molecular weight of 500,000 daltons. A number of subunits are associated in a ring-like structure.

11-3. Beginning of bacterial conjugation. *Escherichia coli,* ×65,000. (J. Gross and L. Caro.)
In some bacterial species, a process of conjugation leads to DNA transfer from a donor cell into a recipient cell. In this picture, the donor bacterium has an elongated form while the recipient is elliptical. A cytoplasmic bridge connects the two, through which the transferred genetic material passes.

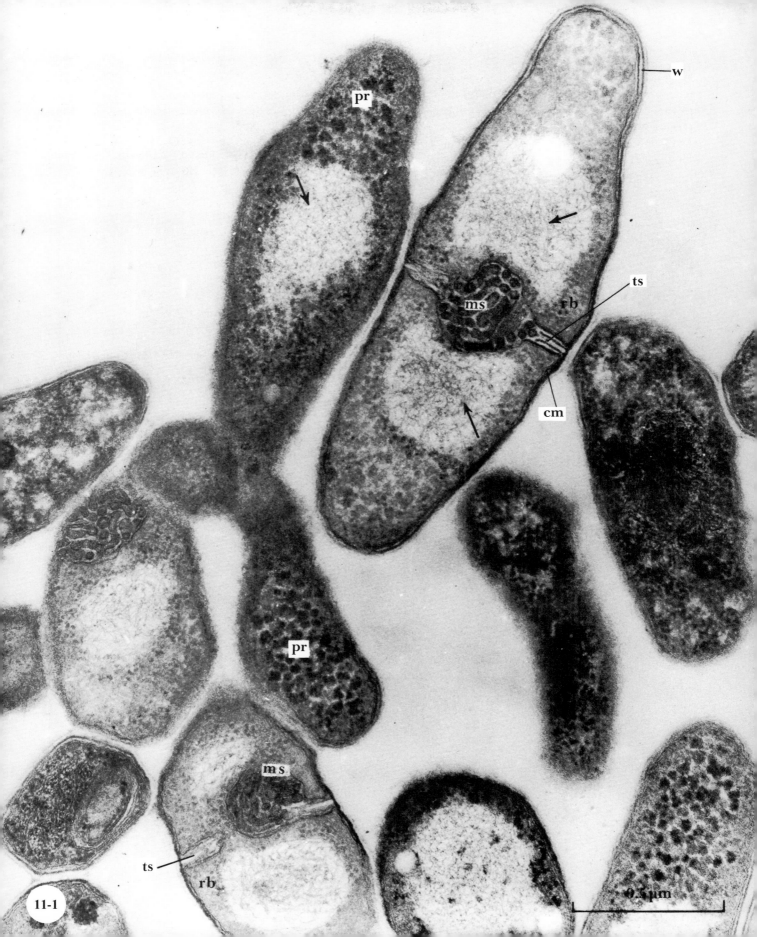

11-1

w

pr

ts

rb

cm

ms

pr

ms

ts

rb

0.5 µm

GENETIC MATERIAL OF BACTERIA: TRANSCRIPTION AND TRANSLATION

Bacteria possess a single circular chromosome that is not enclosed in a nuclear envelope; therefore transcription and translation occur simultaneously. Figures 11-4 and 11-5 demonstrate the function of a bacterial gene. (O. Miller and B. Hamkalo.)

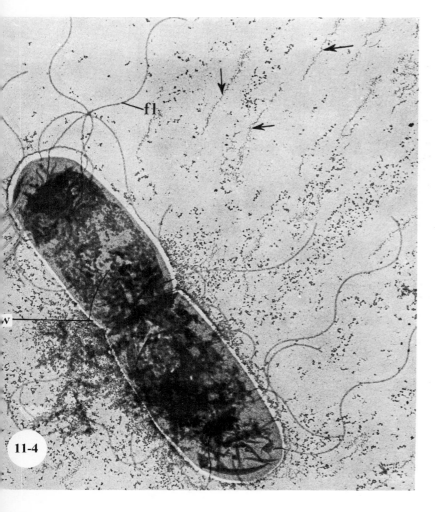

11-4. **A bacterial cell (salmonella) ruptured by osmotic shock.** ×22,000.

Portions of bacterial chromosomes are spread around the ruptured cell. Fibrillar segments are the active ribosomal genes (*arrows*), and the chains of granules are polyribosomes. fl = flagellum; w = wall.

11-5. **An active gene of a bacterial chromosome.** *Escherichia coli,* ×72,000. Ribosomes are attached to molecules of messenger RNA even during the course of transcription. They form polyribosomes (ps) where the genetic information is translated at once into proteins. The granule indicated by the *arrow* is RNA polymerase. This enzyme travels along the DNA and facilitates the progressive synthesis of mRNA.

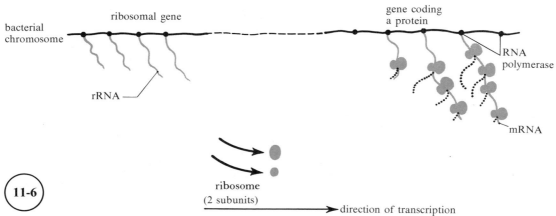

11-6. **Transcription and translation in a bacterial cell.**

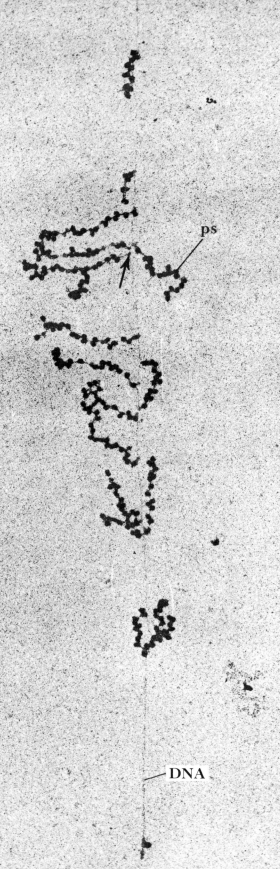

ps

DNA

11-5

0.5 µm

MYCOPLASMAS AND BLUE-GREEN ALGAE

Mycoplasmas do not have cell walls. They have a cytoplasmic membrane which encloses hundreds of ribosomes and a coil of DNA. These cells, which are among the most primitive and simple known, are nevertheless able to synthesize for themselves the proteins and ATP necessary for their metabolism. The blue-green algae are the oldest known organisms. They occur in geological formations, the age of which is estimated as being close to 3 billion years, yet they have a cellular form that is surprisingly complex. In particular, they possess internal membrane systems containing assimilatory pigments that can be considered primitive thylakoids.

11-7. **Mycoplasma structure.** Cells cultivated in vitro, ×68,000. (M. Le Normand, J. P. Gourret, and P. Maillet.) The *arrows* indicate DNA fibers. cm = cytoplasmic membrane; rb = ribosomes.

11-8. **Structure and division of a blue-green alga.** *Oscillaria,* ×24,000. (J. C. Thomas.) Division may be said to occur by the progression of the cell wall as a diaphragm. The daughter cells remain joined and form a filament. Septa of ensuing divisions are already visible (ts$_2$ and ts$_3$). w = wall; cm = cytoplasmic membrane; th = thylakoid; gl = glycogen; sg = secretory granules in the cytoplasm; ts$_1$ = transverse septum.

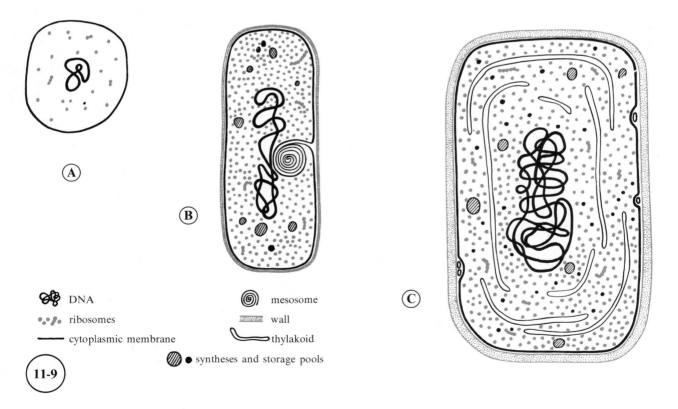

11-9. **Comparison of procaryote structure.**
(A) Mycoplasmas, (B) bacteria, and (C) blue-green algae.

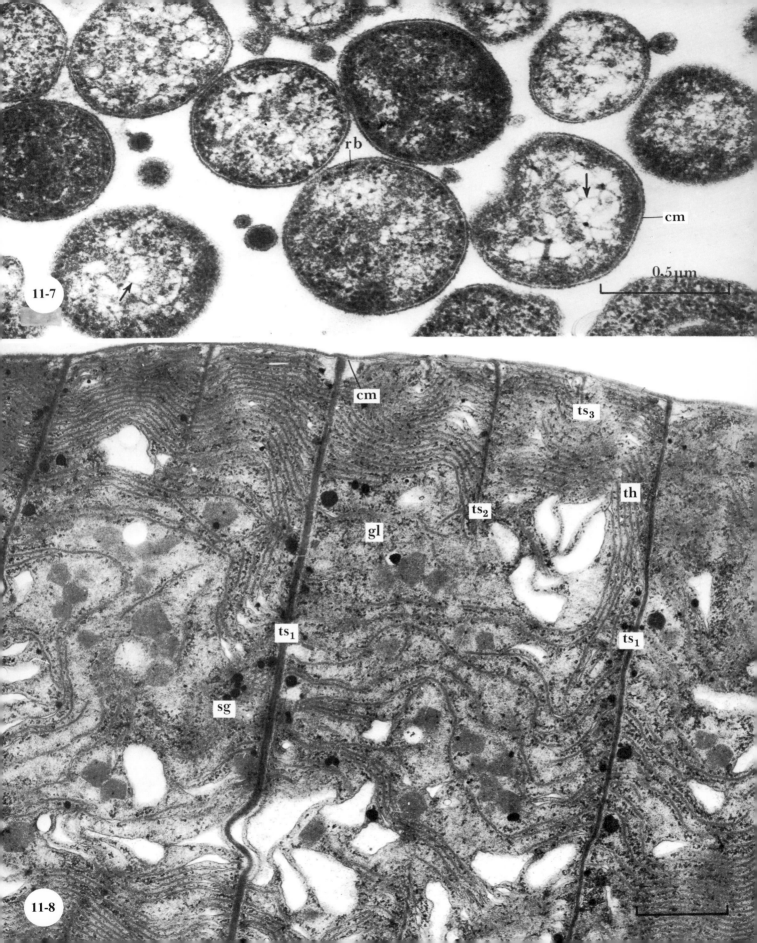

11-7

rb

cm

0.5 μm

11-8

cm

ts₃

th

gl

ts₂

ts₁

ts₁

sg

12. VIRUS

Virus particles, or virions, do not have the organelles described in the preceding chapters and do not have a metabolism of their own. Their reproduction of necessity occurs in living cells, the hosts. Viruses are obligatory parasites. They are constituted of only one nucleic acid, either DNA or RNA. This is protected by a protein shell, called the capsid, during the extracellular phase of the viral life cycle. Certain viruses such as the influenza virus have an additional envelope in which there are lipoproteins and polysaccharides derived from the modified cytoplasmic membrane of the parasitized cells. The nucleic acid of the free viral particle is introduced into the cell. One observes neither replication nor transcription at this stage. Generally, after penetration into the host cell, the nucleic acid replicates by converting to its own use a part of the cellular metabolism. Such a cell is termed infected by an infectious virus. In certain cases, the viral nucleic acid can become integrated with the genome of the host cell. In eucaryotic cells, transformation may then occur; the virus is then termed oncogenic. In bacteria, the integrated virus becomes the provirus; there is lysogeny. The capsid is formed of units, the arrangement of which is the basis for classifying the three types of virus: viruses having helical symmetry; viruses having cubic symmetry; and complex viruses (phage type). In each group the nucleic acid may be either RNA or DNA.

VIRUSES HAVING HELICAL SYMMETRY: TOBACCO MOSAIC VIRUS

This virus parasitizes cultivated plants and has been extensively studied. Its nucleic acid is RNA. The structural protein units of the capsid are arranged in a helix in a very precise manner ($16\frac{1}{3}$ units per turn of the helix), forming a hollow tube. The nucleic acid is lodged in a groove on the internal face of this capsid helix.

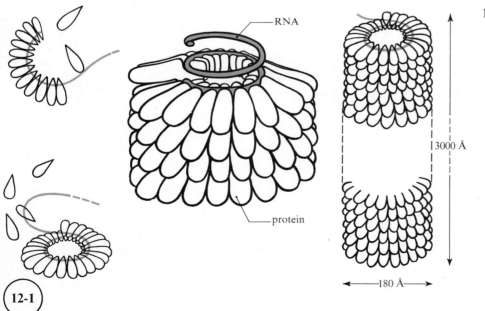

RNA

protein

3000 Å

180 Å

12-1

12-1. **Structure of tobacco mosaic virus (TMV) and self-assembly of the particles from its protein subunits and its molecule of RNA.**
The RNA chain consists of 6600 nucleotides complexed with 2200 molecules of protein.

12-2. **Leaf cell parasitized by TMV.** ×6000. (H. Arnott et al.)
Bundles of viral particles are in the hyaloplasm (*arrows*). One of the chloroplasts (cpl) is abnormal. One of the first signs of parasitic infection is a streaking of the leaves due to a fault in plastid pigmentation.

12-3. **Negative staining of TMV.** ×100,000.
TMV particles are hollow rods (*arrows*) that are striated transversely.

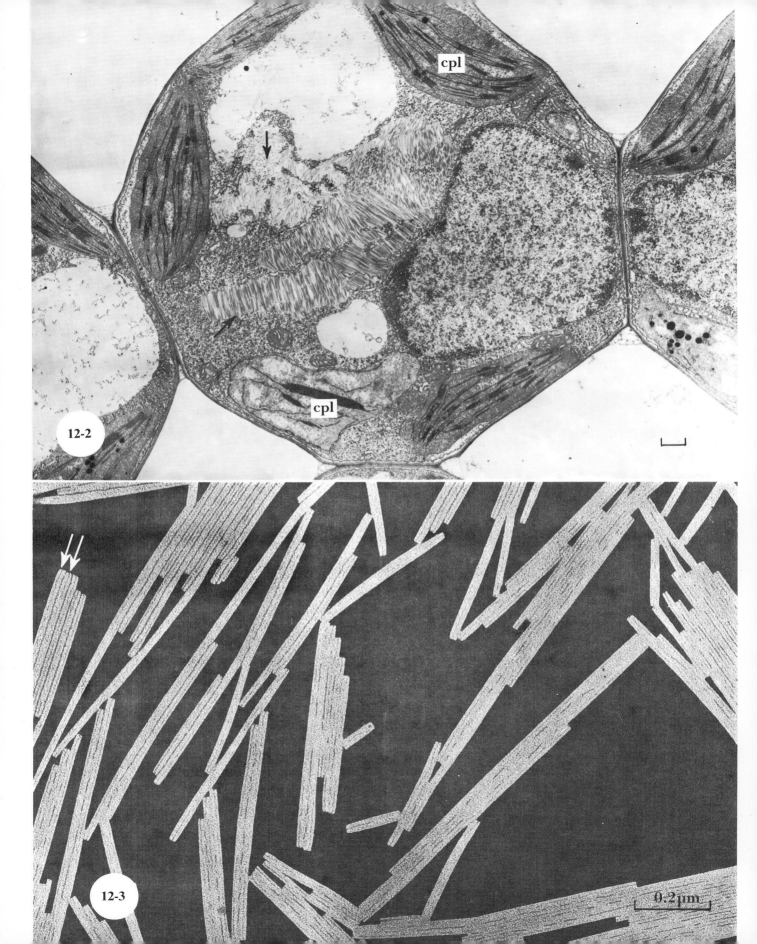

12-2

cpl

cpl

12-3

0·2μm

VIRUSES WITH CUBIC SYMMETRY

The capsid is composed of capsomeres assembled to have icosahedral symmetry (polyhedron having 20 triangular faces). The capsid contains either DNA, as in iridovirus, adenovirus, poxvirus, polyoma virus; or RNA, as in poliovirus.

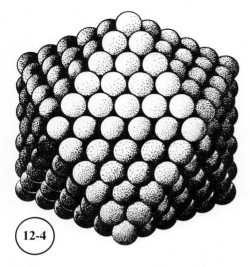

12-4. **Structure of a virus having cubic symmetry.**
The capsid surrounding the nucleic acid is a polyhedron having 12 apexes, the 20 equal faces of which are equilateral triangles. The component capsomeres are each made up of from five to six structural units that are protein in nature.

12-5. **Nereid cell parasitized by an iridovirus (DNA virus).** ×30,000. (G. Devauchelle.)
The virus has multiplied, and the particles have overrun large cytoplasmic areas. m = mitochondria. The significance of the granules localized at the periphery of the viral arrays is not known.

Figures 12-6 and 12-7 show the blood cell (granulocyte) of an insect parasitized by a poxvirus (DNA virus). Penetration of the virus into the cell occurs through phagocytosis. The virions formed after multiplication become coupled with the cytoplasmic membrane and are liberated in the external medium by budding, without immediate lysis of the cell. (G. Devauchelle, M. Bergoin, and C. Vago.)

12-6. **Infected cell.** ×11,000.
Virions (v) are at various stages of development. The *arrows* indicate the particles which are being expelled. n = nucleus.

12-7. **Detail of a viral bud.** ×95,000.
The cytoplasmic membrane (cm) of the host cell forms a viral envelope around the capsid. The *arrow* indicates the nucleoid (DNA) surrounded by proteins.

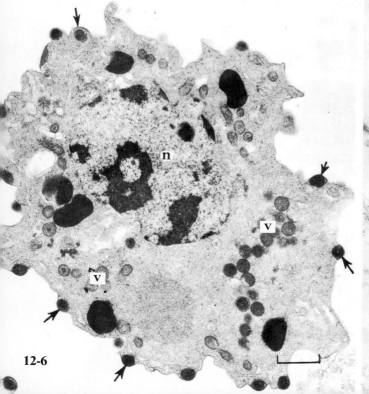

12-6

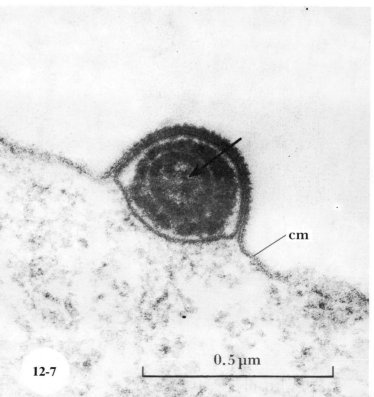

12-7   0.5 µm

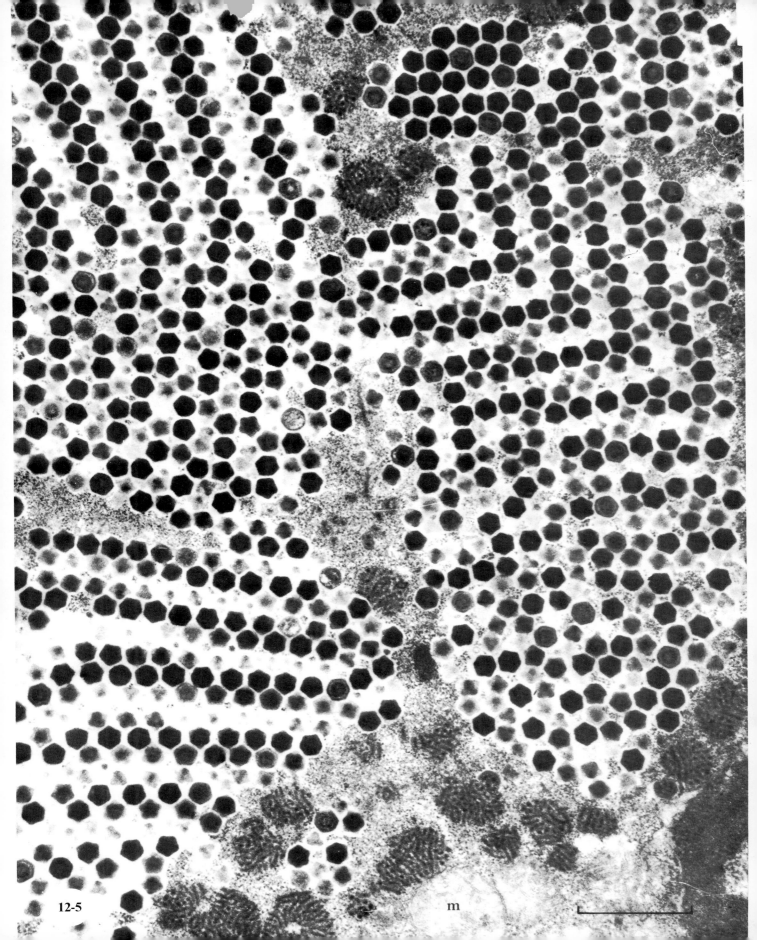

12-5   m

BACTERIOPHAGES

Bacteriophages, or phages, are viruses of bacteria. They are composed of a head having cubic symmetry and a hollow tail having helical symmetry. At the end of the tail there is a fixation system which ensures contact with the parasitized cell. The phage nucleic acid, most often DNA, is contained in the head. It is injected into bacteria through the tail, while the protein capsid remains on the external surface.

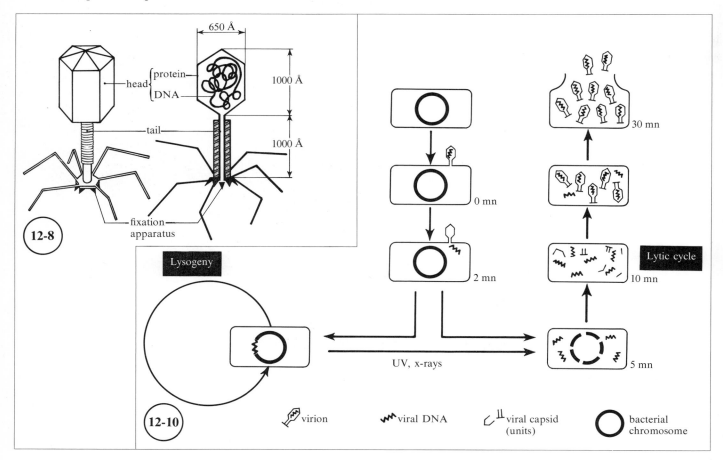

12-8. Bacteriophage structure.
The contractile sheath surrounding the central canal of the tail is involved in injection of the genetic material into the bacteria.

12-9. Streptococcal bacteriophage T_4. Negative staining, $\times 240,000$. (C. Gordon.)
One may distinguish the polygonal head, the striated tail, and the six fibers of the attachment apparatus.

12-10. Bacteriophage cycle.
The cycle consists of attachment to the bacterial wall; injection of viral DNA, which provokes the disintegration of the bacterial chromosome and induces synthesis of enzymes necessary for its own replication; synthesis of viral proteins; DNA capsid assembly; and, finally, bacterial lysis and release of the new viral particles. In certain cases lysogeny occurs: viral DNA is integrated into the bacterial genome without replication and without causing bacterial cell lysis. Such bacteria continue to divide. After several generations and under the influence of external agents such as radiations or chemical substances, the viral genome may reacquire its lytic activity.

12-11. Bacteriophage attached to the bacterial cell wall. $\times 60,000$.
The genetic material has penetrated into the bacterium, and the capsids attached to the bacterial cell wall are empty.

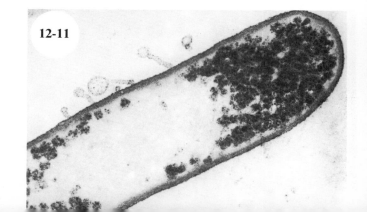
12-11

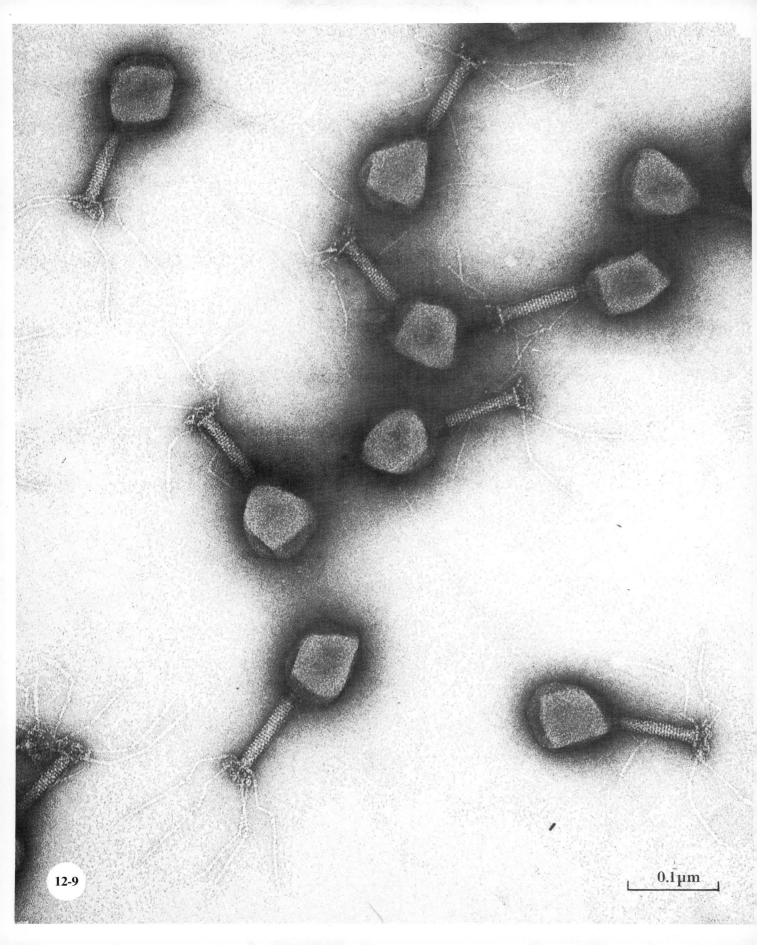

12-9

0.1μm

SUGGESTED READING

ESSENTIAL READING

Ambrose, E. and Easty, D. *Cell Biology.* Reading, Mass.: Addison-Wesley, 1970.
DeRobertis, E. D., Saez, F. and DeRobertis, E. M. *Cell Biology.* Philadelphia: Saunders, 1975.
Dyson, R. *Cell Biology.* Boston: Allyn and Bacon, 1973.
Giese, A. *Cell Physiology.* Philadelphia: Saunders, 1973.
Loewy, A. and Siekevitz, P. *Cell Structure and Function.* New York: Holt, Rinehart and Winston, 1969.
Markert, C. and Ursprung, H. *Developmental Genetics.* Englewood Cliffs, N.J.: Prentice-Hall, 1971.
Novikoff, A. and Holtzman, E. *Cell and Organelles.* New York: Holt, Rinehart and Winston, 1970.
Watson, J. *Molecular Biology of the Gene.* New York: Benjamin, 1976.

GENERAL TREATISES

Dupraw, E. *DNA and Chromosomes.* New York: Holt, Rinehart and Winston, 1970.
Lehninger, A. *Biochemistry: The Molecular Basis of Cell Structure and Function,* 2nd ed. Worth, 1975.
Lima de Faria, A. *Handbook of Molecular Cytology.* New York: American Elsevier, 1970.
Reinert, J. and Ursprung, H. *Origin and Continuity of Cell Organelles.* New York: Springer-Verlag, 1971.
Rothfield, L. *Structure and Function of Biological Membranes.* New York: Academic, 1971.

SURVEY OF METHODS

The reader has doubtless concluded that several methods are needed to explore a structure and to understand it completely. The record, here a photographic plate, represents the culmination of a chosen experimental protocol that has been set up and followed for a defined end. Thus the record can only furnish partial information.

Beyond simple observation, the reader must apply a critical interpretation, taking into account:

1. The purpose of the research (why?)
2. The experimental conditions used (how?)
3. The order of accuracy of the results obtained (significance?)

Before undertaking research of a physiologic or chemical nature, the research worker must know the biologic material to be used and must do a structural study. Aesthetic aspects alone must not hold our attention, because they do not necessarily mean that structures are well preserved. For example, the pretty pictures obtained after fixation with potassium permanganate were found to be the result of a considerable extraction of cellular material, and this fixing agent is no longer used except in special cases.

Cytochemistry and cytoenzymology add to the structural data. By using radioactive tracers which can be followed in space and in time (autoradiography), the static character of ultrastructural records can be compensated for. Finally, direct identification of fractions by microscopy allows us to refine data furnished by biochemists.

The paths of actual research are thus given direction not only by methods of preparation but also by the instruments used to make the observations.

STRUCTURAL INVESTIGATION

Negative staining

This is used for small specimens: macromolecules, viruses, and isolated organelles. It is one of the oldest methods in use. It gives rapid results and is especially suitable for use as a first control of purity for cellular fractions. Fine details such as subunits or periodicity may be demonstrated.

Method: Samples are suspended in a solution opaque to electrons such as phosphotungstic acid. A drop of suspension is deposited on the object carrier. After drying, electron-opaque material is deposited around the object, which appears clear against a dark background.

Shadowing

This technique is applied to viruses and to isolated macromolecules, especially when one dimension is small, such as in the case of fibrils. Shadowing provides highly contrasted images, but the finest details are thickened, and resolution is reduced.

Method: Specimens are deposited on the object carrier, and placed under vacuum in an enclosure where metal is vaporized. Structures retain metallic particles and appear in relief on the object.

Fixing, Embedding, and Sectioning

The purpose of fixing a specimen is to strengthen the cellular structures by forming new intermolecular bonds. It is accomplished by use of chemical substances in buffered aqueous solutions.

Potassium permanganate ($KMnO_4$) preserves membranes but extracts a great many compounds, including nucleic acids. Nucleoli and ribosomes are not preserved. Osmium tetroxide (OsO_4) immobilizes double ethylene bonds, and glutaraldehyde forms bonds between the amine groups of proteins

(R_1 and R_2). Usually a double treatment is used. Glutaraldehyde treatment is followed by a postfixation with OsO_4. Except where noted otherwise (MnO_4K: p. 52, Fig. 5-30A and B; OsO_4: p. 18, Fig. 4-3B; p. 64, Fig. 7-1; and p. 92, Fig. 11-1), double fixation is the method used to prepare the majority of the specimens presented in this Atlas.

Method: In order to be cut in very fine slices (400 Å to 800 Å), the objects are hardened. To this end, they are dehydrated and then impregnated and embedded in plastic. The resulting blocks are cut by precision microtomes armed with a glass or a diamond knife. The sections gathered on the object carrier, usually a fine metallic grid, have insufficient contrast, which must be heightened. They are floated on salt solutions of heavy metals such as uranyl or lead, which fix to the structures and give a positive staining. Ultracryomicrotomes can cut tissues hardened by freezing, thus avoiding dehydration and embedding.

Freeze-etching

This technique has the advantage of avoiding use of a chemical fixative and thus limits the risks of denaturation and extraction. It is applied to many types of samples, but it permits structural studies only.

Method: The material is hardened by rapid freezing to about −150°C by use of liquid nitrogen. Specimen blocks are fractured under vacuum while still frozen and etched by sublimation of the superficial layer of ice. The fracture planes so bared correspond to the planes of least resistance, or the hydrophobic layer within cell membranes. By vaporizing and depositing carbon and metal, a mold or replica is obtained of the fractured surface. The tissues are then dissolved, and it is the replica which is examined.

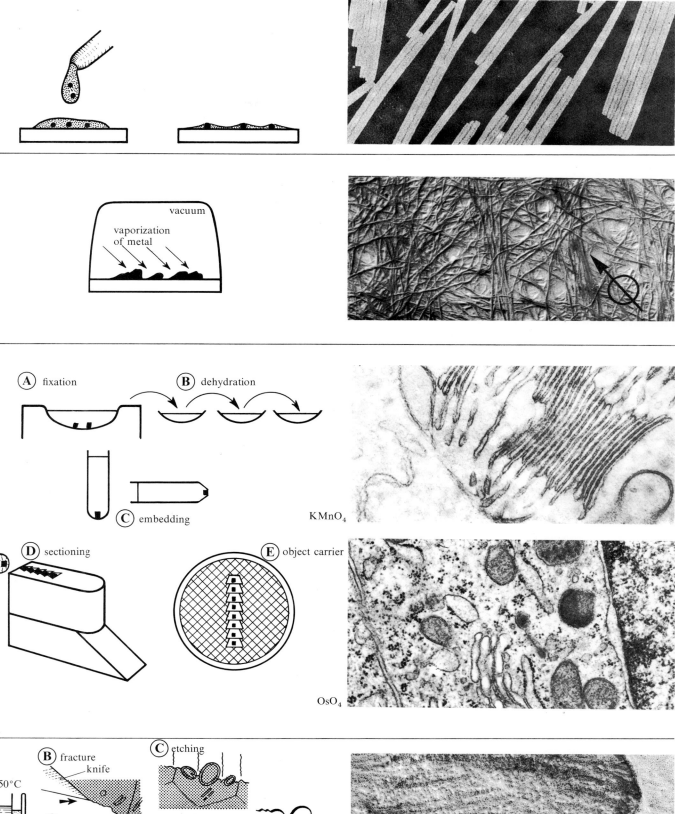

vacuum

vaporization
of metal

(A) fixation   (B) dehydration

(C) embedding

(D) sectioning   (E) object carrier

KMnO₄

OsO₄

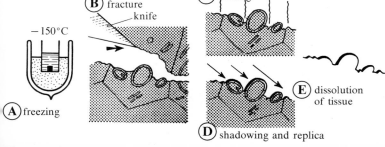

−150°C

(B) fracture
knife

(C) etching

(E) dissolution
of tissue

(A) freezing

(D) shadowing and replica

FUNCTIONAL INVESTIGATION IN SITU

Cytochemistry and selective extractions

Some chemical reactions can be carried out on an ultrastructural scale. Cells are treated by a reagent that combines with certain radicals to give a final product which is opaque to electrons. This treatment can be carried out on blocks or on sections, before or after fixation. The radicals to be studied must be unmasked by a pretreatment at some time. In the detection of polysaccharides, for example, pretreatment consists of a specific oxidation of the alcohol functions of the saccharides by use of a periodic acid (A). The reactivity of the aldehyde groups formed is used to form a complex (B), which is made visible by a silver salt (C). Control reactions must be done to ascertain that the reaction is specific. An indirect method of characterizing a substance consists of its extraction, whether by a specific chemical solvent or by a hydrolytic enzyme such as protease, amylase (p. 56), or lipase.

Cytoenzymology

The presence of an enzyme in a cell is detected by the products of its activity. To identify phosphatases, for example, the tissues are incubated in a medium containing a phosphate, which is the substrate, and a soluble salt of lead. The presence of an enzyme is noted through the liberation of phosphate (A) and by the formation of a lead phosphate precipitate that is opaque to electrons (B).

To identify peroxidases, diaminobenzidine (DAB) in its oxidized form (DABox) is combined with osmium to give an insoluble product opaque to electrons. It is used as the substrate in an oxidation-reduction reaction in which hydrogen peroxide (H_2O_2) is reduced. Cytoenzymologic reactions must be controlled for enzyme denaturation (by heat) by using specific inhibitors, such as fluoride for phosphatases or cyanide for respiratory enzymes, and by carrying out incubations in media without substrate (*in the circle*).

Immunocytochemistry

In order to localize a substance in a cell having antigenic properties (proteins, glycoprotein), one can use the ability the substance has in forming complexes with its antibody. Substance X is injected into an animal such as a rabbit. The elaborated antibodies (anti-X) are withdrawn and purified. The antibody is coupled with a molecule opaque to electrons, such as ferritin, or with an enzyme which is easily detectable, such as peroxidase. In the presence of substance X, the marked antibody forms a visible complex with it or a complex which is demonstrable by electron microscope.

Autoradiography

When a metabolite containing a radioactive atom is given to a living cell, the labeled metabolite is used like the nonradioactive isotope for the cell's syntheses. The atom is found incorporated in the natural constituents of the cell and serves as a tracer. Precursors marked with ^{3}H are used most often. The molecule chosen to bear this atom is one which will be specifically incorporated in a macromolecular structure: for example, ^{3}H-thymidine for DNA; ^{3}H-uridine for RNA; amino acids for proteins; monosaccharides for polysaccharides.

Soluble compounds are eliminated by treatment. Detection of the incorporated activity is made by applying a photographic emulsion to the section. The rays emitted by the disintegrating radioactive atoms leave an image on the emulsion which is eventually revealed in the same way as an ordinary photographic film. The metallic grains of silver are localized above the structures which incorporated the radioactive atoms.

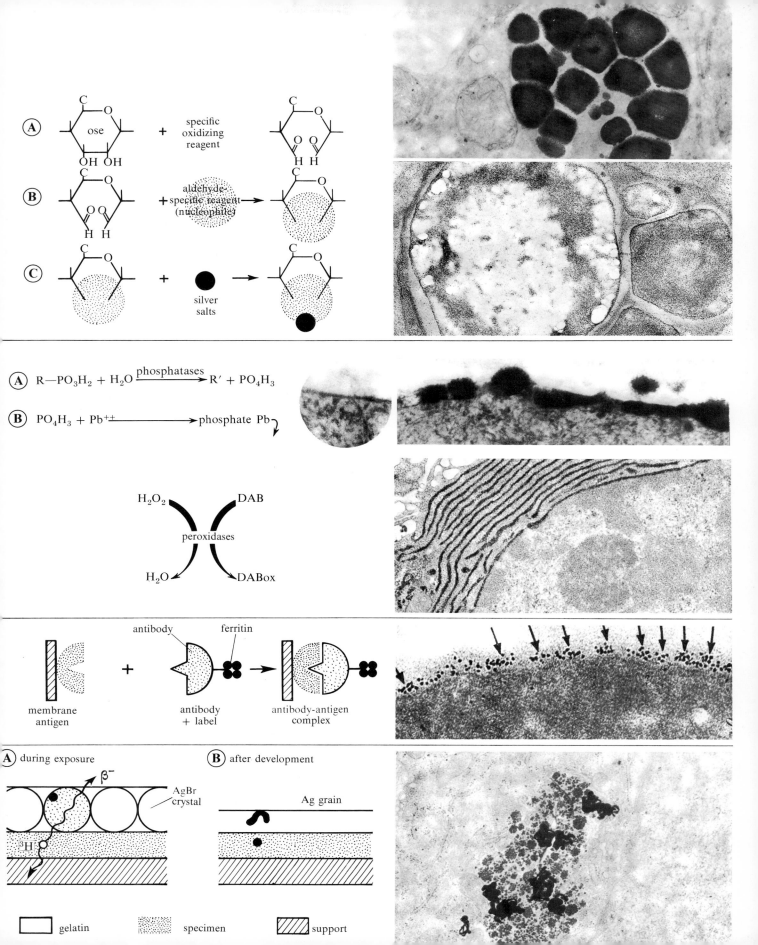

A ose + specific oxidizing reagent

B (aldehyde) + aldehyde-specific reagent (nucleophile) →

C + silver salts →

A $R-PO_3H_2 + H_2O \xrightarrow{phosphatases} R' + PO_4H_3$

B $PO_4H_3 + Pb^{++} \longrightarrow$ phosphate Pb

H_2O_2 + DAB → peroxidases → H_2O + DABox

membrane antigen + antibody + label (antibody + ferritin) → antibody-antigen complex

A during exposure

β^-

AgBr crystal

$^3\underline{H}$

B after development

Ag grain

gelatin   specimen   support

Electron microscopes

When one speaks of the electron microscope, one generally is referring to the transmission microscope, which is an apparatus in which the specimen is traversed by a beam of electrons. Since the 1960s, this instrument has permitted considerable progress in our knowledge of cellular structure. Thanks to refinement of techniques, its field of action extends to the study of macromolecules, which are the functional basis of the living cell. At the same time that methods have been diversifying, new types of microscopes were being constructed. The scanning electron microscope, microprobe, high-voltage microscope, and other apparatuses are being studied. Their use promises many more new developments of great importance for cellular biology and molecular biology.

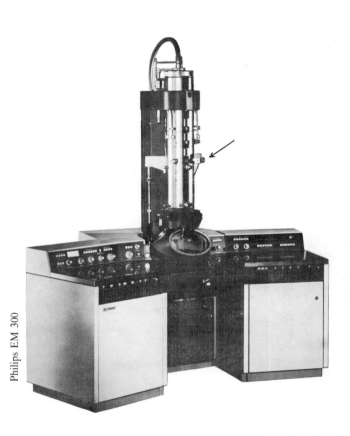

Philips EM 300

Transmission microscope

In principle this microscope is similar to the light microscope, but instead of being composed of photons, the beam used is composed of electrons, the wavelength of which is on the order of an Angstrom unit.

Electrons are accelerated by a potential difference of 60,000 to 100,000 volts, and the beam is focused by two electromagnetic lenses. The image is displayed on a fluorescent screen, and the area of interest is examined by a binocular magnifier. The image is recorded on photographic plates placed under the screen.

The specimen is held on the object carrier and is introduced into the column (*arrow*). The very high vacuum needed for the electron beam prevents observations on living cells.

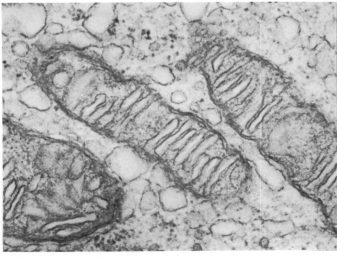

Scanning electron microscope

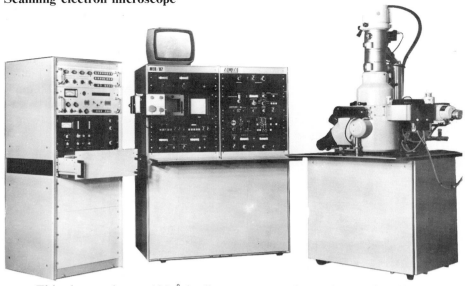

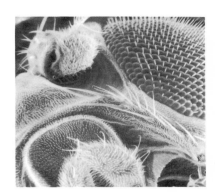

In this apparatus the electron beam is reflected by the specimen. This electron beam, 100 Å in diameter, scans the surface of the object; the image forms in a cathode tube from the reflected electrons. Acceleration voltages used are relatively weak (20,000 to 30,000 volts) so that there will not be electron penetration into the specimen. Resolution is generally in the order of 100 Å. The principal advantage of this microscope is its great depth of field and its capacity to explore large surfaces. Its domain of usefulness was for some time limited to studying hard objects, but careful methods of dehydration now have made it possible to examine fragile specimens such as leukocytes and Protista.

High-voltage microscope

Electrons are accelerated by a very high voltage (1 to 3 million volts) in this microscope. As a result, penetrating power into an object is considerably increased, permitting observation of thick biological objects (several μm) and exploration of a cellular volume.

This penetrating power also permits the study of living specimens in microchambers sealed against the vacuum. Bacteria have been maintained in these enclosures under normal living conditions of pressure and humidity. Thus protected, it has been possible to observe them alive during a brief period of time in the high-voltage microscope of Toulouse using a potential difference of 1 million volts.

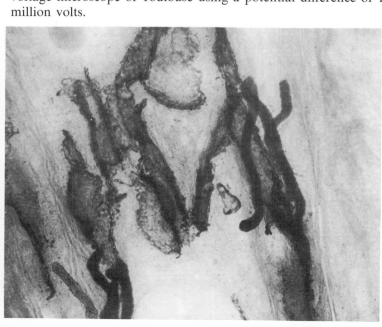

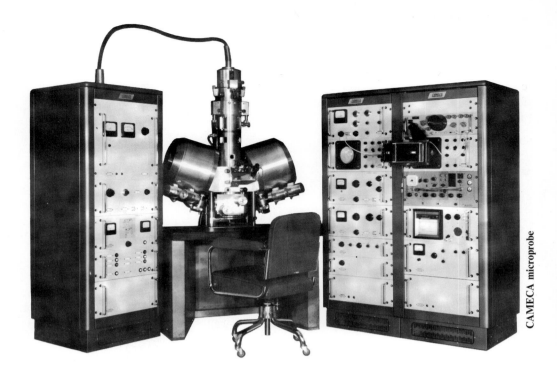

CAMECA microprobe

Electron microprobe

When hit by an electron beam, objects emit x-rays having a wavelength characteristic of the atoms bombarded. Spectrographs measure the wavelengths of the emitted x-rays and lead to a determination of the elemental chemical composition of extremely small volumes of substances (on the order of $1 \mu^3$). Thus it is possible to carry out structural examination and chemical analysis of a specimen at the same time.

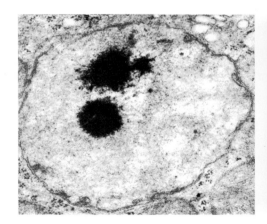

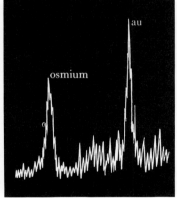

ABBREVIATIONS

a	amylaceous (starch-bearing) vesicle		m	mitochondrion
ac	acrosome		ms	mesosome
af	axial fiber		mt	microtubule
au	gold		mv	microvilli
			mx	mitochondrial matrix
bb	basal body		my	myofibril
bc	bile canaliculus			
br	Balbiani's rings (puffs)		n	nucleus
			np	nuclear pore
c	cilia		nu	nucleolus
C	cytochrome C			
C_1, C_2	a pair of centrioles		phv	phagocytic vacuole
ch	chromatin		pl	plastid
chr	chromosome		pm	plasmodesma
cm	cytoplasmic membrane (plasmalemma)		pns	perinuclear space
cpl	chloroplast		pr	polysaccharide reserve
cs	centrioles		ps	polyribosome
			px	peroxisome
d	dictyosome			
			rb	ribosome
en	nuclear envelope		rer	rough endoplasmic reticulum
enm	external nuclear membrane			
er	endoplasmic reticulum		s	plastid stroma
			ser	smooth endoplasmic reticulum
fl	flagellum		sg	secretory granule
g	Golgi apparatus		th	thylakoids
gl	glycogen		ts	transverse septum
gr	chloroplast granum		tv	transition vesicle
gx	glycocalyx			
			um	unit matrix (gene)
icr	internal crista of mitochondrion			
inm	internal nuclear membrane		v	virus
l	lipid		w	wall
lm	middle lamella			
ly	lysosome			

ADP	adenosine diphosphate
ATP	adenosine triphosphate
DNA	deoxyribonucleic acid
FAD	flavin-adenine-dinucleotide
NAD	nicotinamide-adenine-dinucleotide
NADP	nicotinamide-adenine-dinucleotide-phosphate
mRNA	messenger RNA
rRNA	ribosomal RNA
tRNA	transfer RNA

INDEX